Chemistry and Mode of Action
of Crop Protection Agents

Chemistry and Mode of Action of Crop Protection Agents

Leonard G. Copping
LGC Consultants, Saffron Walden, Essex, UK

H. Geoffrey Hewitt
University of Reading, UK

THE ROYAL
SOCIETY OF
CHEMISTRY
Information
Services

A catalogue record for this book is available from the British Library.

ISBN 0-85404-559-7

Published by The Royal Society of Chemistry,
Thomas Graham House, Science Park, Milton Road, Cambridge CB4 4WF, UK
For further information see our web site at www.rsc.org

Typeset by Paston Press Ltd., Loddon, Norfolk
Printed by Redwood Books Ltd., Trowbridge, Wiltshire

Preface

In the time it takes to read this sentence another twenty people will have been added to the world's population and by this time next week enough people will have been born to establish a new city about the size of Birmingham in the UK or Detroit in the USA or Brisbane in Australia. This rapid expansion is forecast to stabilise by 2050 at 11 billion, a 100% increase from 1998. Enormous sociological and economic progress must occur to allow such increases without apocalyptic penalty but of primary importance will be our ability to produce food in sufficient quantity and of appropriate quality to sustain an acceptable standard of living.

Unfortunately, we start in the weak position of already having a global food shortage. This fact may not be as close to those of us fortunate enough to live in the more developed parts of the world, yet the all too frequent incidences of famine in developing regions should remind us that, despite our current level of political and technological expertise, a large proportion of the population lives on a knife-edge of food insecurity. Paradoxically, in the developed world where modern farming methods are well established and support industrialisation and a high standard of living together with high arable productivity, some people complain about an excess of food and seek to return to traditional practices. Clearly, in a world of plenty there are many problems but in a world of famine there is only one.

Increased food production can be achieved either through expanding the area used for agriculture or by increasing levels of productivity per unit area. Essentially, the more primitive the farming, the more extensive it has to be to sustain an adequate food supply, compared with high input farming methods that optimise land use and yield. However, land is a finite resource and the needs of a growing population are just one factor in balancing land development and land conservation. There is general agreement that, on a global scale, the increasing population can only be fed through intensive farming. Indeed, there is an excellent argument that such high input farming is more beneficial to conservation than traditional methods in that it allows less productive land to be left for biodiversification whilst the high yielding land is used to produce food.

Crop protection agents ('pesticides' or 'agrochemicals'), play an essential role in the spectrum of activities that comprise modern farming. Every crop is attacked and antagonised by a characteristic array of weeds, insects and fungi, and, without crop protection agents, an estimated 30% of yield would be lost and more would be consumed or spoilt by storage pests and pathogens. Not all crops benefit from the use of crop protection agents. Where water shortage or nutrient levels are severe, yield is not limited by the incidence of pests, diseases or weeds and to apply pesticides would be foolish. For example, early cereal yield increases were directly related to the amount of applied nutrient, particularly nitrogen fertiliser, and only when nutrient deficiencies diminished did it become necessary to combat the associated growth of weeds with herbicides in order to preserve the inherent yielding capacity of the crop. In the US wheat belt, extensive cereal cultivation would benefit more from increased water availability than from fungicide use. However, in high yielding cereal monocultures, high plant nitrogen levels increase the susceptibility of the crop to fungal attacks of epidemic proportions, and, in those situations, crop yield can only be protected through the use of appropriate fungicides.

The use of crop protection agents is not new, although until quite recently the reasons for their activity were not understood. The first agents were effective enough to reveal to the farmer the economic advantages of weed, disease and insect control but, by modern standards, they were crude, being applied at high rates and lacking the spectrum, selectivity and reliability that are essential attributes in modern products. These evolved largely through serendipity working closely with good observational science, relying more upon the chance discovery of commercially useful activity than on a rational approach to design. Yet, as more and more products were announced, each a little better than the competition, relationships between structure and activity emerged and biochemical modes of action were determined. Each generation of product offered a new advance. For example, herbicide rates of application fell from several kilograms per hectare to less than 10 g.

There are few important weed, insect or disease problems in major crops that have not been addressed and for which there is not at least one synthetic crop protection agent available to the grower. The challenges to the agrochemical industry are now associated with the search for novel, competitive products in an already crowded market, for which use strategies can be developed to combat the spectre of resistance and which comply with an ever increasing legislation that demands safety to the user, environment and consumer. Essentially, the problem is one of cost. Escalating research and development expense now means that an investment of about $200 million is needed to launch every new product. In that economic environment, only the large companies can survive and even they are exploring other means to achieve the success currently enjoyed by synthetic crop protection agents.

Although considerable effort is still directed towards the discovery of new synthetic molecules developed empirically or through rational design, the natural world is now seen as a rich source of novel chemistry and mode of action. Increasing emphasis is placed upon the discovery of activity in natural products or secondary metabolites that can be modified chemically to produce patentable and useful crop protection agents. The biochemistry of plants is being exploited to manufacture natural defence products in response to applied synthetic triggers and, similarly, the expression of useful characteristics in crops through the incorporation of (unusual) foreign genes coding for proteins that give the transgenic crop advantages over conventional varieties is providing new opportunities for pest control. Such developments are proceeding in concert with the philosophy of integrated crop management in which the means of weed, insect and disease control are optimised, according to the situation, and form just one facet of a holistic approach to modern farming.

There can be no doubt that the discovery, development and use of crop protection agents is one of the greatest success stories in the 20th century. Through the directed application of novel chemistry towards meeting biological goals, they help to feed the world. The nature of crop protection is changing but the continued partnership between the agrochemical industry and farming must be defended if we are to survive.

This book describes the chemistry and mode of action of the three major groups of crop protection agents, herbicides, insecticides and fungicides. A short section also highlights the use of plant growth regulators as synthetic materials designed to interact with crop growth and maximise commercial yield. It is designed as a general introduction to those studying agricultural science, to advisers, consultants, academics and industrialists. Each chapter is fully referenced thereby allowing the reader to pursue areas of special interest and there is, wherever relevant, an attempt to explore the future direction of an exciting and essential science.

Leonard G. Copping and H. Geoffrey Hewitt

Contents

CHAPTER 1

Introduction

Crop protection is big business. It is big for farmers, it is big for manufacturers and it is big for the research-based groups involved in agrochemical research. However, there is a general perception amongst the general public that pesticides are bad. The argument follows the lines of: look at the effects of DDT; consider the damage caused by Agent Orange in Vietnam; and how many people have been affected by agrochemicals whilst innocently walking in the countryside. Anything that is that destructive has to be all bad.

Is this a fair reflection of the contribution that crop protection chemicals have made to the world in which we live? Are there no benefits at all? Why do we accept the continual pollution of our land, the presence of toxic residues in our food and the destruction of all non-target organisms by the indiscriminate over-application of chemicals?

It was interesting to hear Steve Jones, Professor of Genetics at University College, London, state on the Food Programme on Radio 4, that the population of the world is increasing at an alarming rate and has been doing so for some 20 or 30 years. Nevertheless, the amount of food consumed by each individual on earth, not just the total amount of food, has increased over that same time period. So each person living today has more to eat than their parents had thirty or even twenty years ago. Clearly we are getting something right.

The main problem is that it is bad news that people report. The fact that DDT saved millions of lives at the end of the Second World War and into the 1950s and 1960s is not news but the fact that it is persistent and accumulates in the food chain, causing the death of birds, is. I am sure that I am not alone when I say that I would rather have a small residue of DDT in my body than suffer from malaria. Today we have choices in the insecticides that we can use. In 1945 we did not.

Similarly, we are all appalled at the tragic loss of life following the accident at Bhopal, India. However, the item that was top of the newspaper reporting of the incident was that the factory made an

1

insecticide. Not the tragedy of the incident; not the possibility that quality assurance and quality control at the factory were lax; not that there was poor supervision of a process that, if allowed to run out of control, could lead to an explosion; not that this was a particularly tragic industrial accident. The factory was making an insecticide – a poisonous chemical designed to kill – and this was the main theme adopted by reporters.

How many reports are there that, in the UK, yields of wheat per hectare have risen significantly over the last twenty years? It is not unusual for yields of around ten tonnes of grain per hectare to be harvested from winter wheat crops. The direct effects of agrochemicals to the farmer range from between three and five times the value invested (Pimentel *et al.*, 1991;[1] LeBaron[2]). World Health Organisation (WHO) reports indicate that the use of pesticides has made a significant contribution to farming practice by reducing labour requirements, conserving fossil fuels, increasing crop yields, lowering food costs and improving food quality (Borlaug;[3] World Health Organisation;[4] National Research Council Board of Agriculture;[5] Smith *et al.*;[6] Sweet *et al.*;[7] LeBaron[2]).

Recent studies in the US have indicated that if crop protection chemicals were banned yields of fruit, vegetables and cereal crops would decline by 32 to 78% (Smith *et al.*[6]). What effect would this have on the price of food?

1 FUTURE NEEDS

It is estimated that there will be an additional 3 billion people to feed in the world by 2025 and, by 2050, the population is expected to exceed 11 billion, more than twice today's population.[8] This means that within the next 50 years it will be necessary to produce more than twice as much food as is currently being produced.[3] It must always be remembered that if the population increases, the land available for agricultural production will fall as these new people will have to live somewhere. Today, the amount of arable land available for the production of food per person is down from the half-a-hectare figure of the 1960s to about one-third of a hectare.[9] Each available hectare must support more and more people as world population continues to increase at a rate of 1.7% per year (90 million more people to feed and clothe each year), whilst the rate of expansion of world cropland is less than one tenth of this rate (0.15% per year or 50 to 60 million new hectares of cropland by 2010).[9] In less than twenty years each person will have to be supported by only 0.2 hectares.

With more people and less land per person, the only way that the population of the world can be fed is to increase productivity per hectare. The only way to do this is with improved crop protection. We will not be able to survive without the strengths of science and technology and the judicious application of crop protection agents.

2 WORLD MARKET IN 1995

Global agrochemical sales rose by 11.9% to US$28 965 million at end-user level in 1995. When this figure is discounted for the effects of inflation and currency fluctuations, growth, in real terms, is estimated to be 4.3% over 1994. This is the second year where real increases in sales value have been recorded (1994 sales were 5.1% higher than 1993 figures).[9]

The majority of these agrochemical sales are controlled by the 25 companies whose annual income from agrochemicals represent over 90% of the pesticide market (Table 1.1).

These data tell us that to be a successful player in the pesticide industry you must be a large organisation with the ability to invest a great deal of support into the discovery (in most cases), development, manufacture and marketing of your products. There are some organisations that have built their successful position on their ability to manufacture and formulate commodity products, those products that were discovered some years ago and whose patents have lapsed, thereby allowing organisations, other than the inventor to make, formulate and sell the product internationally.

Most of the top 25 companies are involved in discovery research targeted at the synthesis of new chemicals with new chemical structures, new modes of action and low rates of application that can be protected by international patents and that will give the inventing company an advantage over its competitors. Other generic manufacturers also invest in research but this is more applied in terms of manufacturing opportunities and formulation advances.

The money that is invested in research and development is thought to be a reflection of the chances of successful discovery. Without a doubt, the more money that is invested, the more research that can be done. If, however, the investment is say 10% of sales income and the growth of the market is 5% per annum, then the company is investing too much as it will not recoup its investment.

The most money is being invested by the companies with the largest sales. In broad bands the money being invested is:

Table 1.1 *Agrochemical sales of the leading 25 companies*

Ranking 1995 (1994)	Company	Agrochemical sales[a]		% Change vs 1994	
		$ millions 1995 (1994)	National Currency	$	National Currency
1 (1)	Ciba	3 284 (2 954)	SwFr3 880 (SwFr4 037)	+ 11.2	− 3.9
2 (3)	Monsanto[b]	2 472 (2 224)	US$2 472 (US$2 224)	+ 11.2	+ 11.2
3 (4)	Zeneca	2 363 (2 105)	£1 497 (£1 374)	+ 12.3	+ 9.0
4 (5)	AgrEvo	2 344 2 045	DM3 358 (DM3 317)	+ 14.6	+ 1.2
5 (6)	Bayer	2 332 (1 950)	DM3 341 (DM3 163)	+ 19.6	+ 5.6
6 (2)	Du Pont	2 322 (2 132)	US$2 322 (US$2 132)	+ 8.9	+ 8.9
7 (7)	Rhone-Poulenc[c]	2 068 (1 804)	FFr 10 313 (FFr 10 005)	+ 14.6	+ 3.1
8 (8)	DowElanco	1 962 (1 735)	US$1 962 (US$1 735)	+ 13.1	+ 13.1
9 (9)	Cyanamid (AHP)	1 910 (1 600)	US$1 910 (US$1 600)	+ 19.4	+ 19.4
10 (10)	BASF	1 432 (1 258)	DM2 052 (DM2 040)	+ 13.8	+ 0.6
11 (11)	Sandoz	1 125 (1 001)	SwFr1 329 (SwFr1 368)	+ 12.4	− 2.9
12 (12)	Sumitomo[d]	635 (603)	¥65 373 (¥60 210)	+ 5.3	+ 8.6
13 (13)	Kumiai[e]	597 (552)	¥56 156 (¥56 388)	+ 8.2	− 0.4
14 (14)	FMC[f]	590 (525)	US$590 (US$525)	+ 12.4	+ 12.4
15 (15)	ISK[g]	535 (465)	¥50 300 (¥47 563)	+ 15.1	+ 5.8
16 (16)	Sankyo[h]	505 (448)	¥47 500 (¥45 800)	+ 12.7	+ 3.7
17 (18)	Rohm and Haas	498 (439)	US$498 (US$439)	+ 13.4	+ 13.4
18 (17)	Nihon Nohyaku[h]	467 (446)	¥43 913 (¥45 542)	+ 4.7	− 3.6
19 (20)	Nissan Chemical[g]	410 (370)	¥38 562 (¥37 816)	+ 10.8	+ 2.0
20 (19)	Hokko[i]	407 (371)	¥38 274 (¥37 904)	+ 9.7	+ 1.0
21 (21)	Takeda[g]	388 (357)	¥36 462 (¥36 470)	+ 8.7	0.0
22 (23)	Makhteshim-Agan[j,k]	378 (331)	US$378 (US$331)	+ 14.2	+ 14.2

(continued)

Table 1.1 *Continued*

Ranking 1995 (1994)	Company	Agrochemical sales[a]		% Change vs 1994	
		$ millions 1995 (1994)	National Currency	$	National Currency
23 (24)	Uniroyal[h]	326 (268)	US$326 (US$268)	+21.6	+21.6
24 (–)	Fernz	285 (174)	NZ$ 435 (NZ$ 294)	+63.8	+48.0
25 (25)	Nippon Soda	245 (201)	¥23 037 (¥20 567)	+21.9	+12.0

[a] Converted using annual average exchange rates; [b] 1995 figure includes undisclosed animal health product sales that were reported separately, at $101 million, in 1994; [c] includes inter-group sales; [d] figure supplied by Sumitomo is for calendar year, although company year end is March 31st and figure is reported in dollars and converted at year end exchange rate quoted by Sumitomo ($1 = ¥99.85 in 1994 and $1 = ¥102.95 in 1995); [e] year ended 31st October; [f] estimate; [g] year ended 31st March; [h] year ended 30th September; [i] 1994 sales restated; [j,k] figures reported in US$.

R&D spending $ millions	Company (approximate % of company sales)
250–300	Ciba (10%);
200–249	AgrEvo (11.5%), Bayer (13%), Zeneca (10.5%);
150–199	BASF (11%), Cyanamid (10%), DowElanco (11%), Du Pont (11%), Rhone-Poulenc (11%);
100–149	Monsanto (7%), Sandoz (11%);
50–99	FMC (12%), ISK Bioscience (11.5%), Sumitomo Chemical (11%);
<49	Hokko (4.2%), Kumiai (5.2%), Fernz (10%), Makhteshim-Agan (9%), Nihon Nohyaku (10%), Nippon Soda (11%), Nissan (10%), Rohm and Haas (10%), Takeda (9%), Sankyo (9%), Uniroyal (10%).

However, it is unlikely that the introduction of new and more sophisticated crop protection systems will assist the developing world. Here there are currently compounds available to solve some of their problems but there is little or no money to supply the necessary products. It is true that non-governmental organisations (NGOs) do aid the agricultural systems of these countries but this aid is often politically motivated and is coloured by the objectives of the country supplying the funding. For example, paraquat was widely used as an effective post-emergence herbicide for the control of all weeds as a post-directed spray within a wide range of perennial crops and as a seed preparation treatment for arable land. It suffers from the problem that if consumed as the commercial product it will often cause irreversible damage to lung tissue. The material is often transported within the poorer countries within inappropriate vessels that may be re-used to carry water or to prepare food. The consequences are obvious. The solution seems to be to ban a valuable product rather than use the expert formulation technology developed within the manufacturers' laboratories to render it unpalatable and difficult to transfer to alternative containers. The improved formulation, of course, costs more but saves lives and also protects the crop.

Another aspect is that agrochemical companies will not develop new compounds to control specific problem weeds, pests and disease in the crops of the developing world. Hence, the staple crops of Africa such as sorghum and cassava will not be within the screening framework of the major players and any development into these crops, and it is essential that all compounds that will be used in these crops are shown to be effective and safe to the consumer, the crop and non-target organisms, will have to be funded by another organisation. Similarly, again in Africa, parasitic weeds such as *Striga* (witchweed) – often parasitising sorghum – and *Orobanche* (broomrape) – often parasitising leguminous crops – cause huge losses, often as high as 75% but there is little work undertaken to examine techniques for their control. However, when *Striga* was found growing in South Carolina, the US government invested a huge amount of time and effort in its eradication and this has now been successfully achieved. The costs of the programme, however, were so high as to be of no benefit to the people who suffer annually from the effects of these weeds.

Rice is a crop that appears in the list of the most important crops in terms of money spent for crop protection. It is also the mainstay of many developing countries. It is fortunate that, in Japan, agriculture is so well-developed that many companies design compounds specifically for the Japanese rice crop. Many of these, but not all, find application in developing countries although this is often after the expiry of their global patent. There is also the International Rice Research Institute (IRRI) in the Philippines, funded by NGOs and supported by charitable organisations such as the Rockefeller Foundation, which strives to improve yields of rice within the poorer developing countries.

The introduction of pest control methods is often a result of the broad-spectrum activity of compounds that have been commercialised in other crop situations. Control of locusts, for example, is an important aspect of crop protection in much of Africa and the Middle and Far East. The compounds that are used are always compounds such as chlorpyrifos that have a major market outlet, that are available because of their market uses and that can be sold in emergency situations to aid agencies for application. It falls to non-commercial groups such as Commonwealth Agricultural Bureau to fund research that is targeted to this type of problem as the primary use.

The number of agrochemical companies is reducing as many merge to form larger, more securely financed companies or are acquired by their larger competitors. Such acquisitions include the take over of Shell Development (USA) by Du Pont and the subsequent acquisition of Shell Research by American Cyanamid (and its subsequent acquisition by American Home Products), Ciba-Geigy's purchase of Maag, Sumitomo's acquisition of Chevron to form the US-based operation Valent and Rhone-Poulenc's acquisition of Union Carbide. Mergers are also well established with the most important being the merger between CIBA and Geigy to form Ciba-Geigy and the subsequent merger with Sandoz to form Novartis (now the largest agrochemical business with predicted annual sales of over $4 billion), Dow and Elanco to form DowElanco, and Hoechst and Schering (who had already purchased the UK-based merged company FBC formed by the collaboration of Boots and Fisons) to form AgrEvo.

2.1 1995 Sales by Category

In the developed world in markets such as small grain cereals, soybean, maize and rice, the value of the herbicide market is much larger than either insecticides or fungicides (Figure 1.1). This is because the econo-

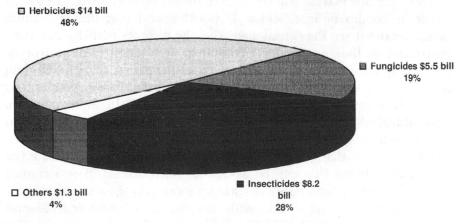

□ Herbicides $14 bill
48%

■ Fungicides $5.5 bill
19%

□ Others $1.3 bill
4%

■ Insecticides $8.2 bill
28%

Figure 1.1 *World agrochemical market sales by category – 1995*

mies of developed countries, *e.g.* in Europe and North America, has led to a move of the population from rural environments where they worked the land, to the urban environment. Weeds can be guaranteed to occur within any intensively farmed area and weeds will always have a deleterious effect on the crop. This effect may be a direct yield loss through competition for light, water or nutrients. It may be an indirect effect on the value of the crop through increased difficulty in harvesting. It may be through the introduction into the crop of poisonous weed by-products with a deleterious effect on the health of the consumer or it may contain weed seeds that will reduce the crop's value. The movement of people from the land removes the labour that was available for hand-weeding and so chemical weed control becomes essential.

The share of the herbicide market of individual companies varies with the success that each has found with its discovery programme. Table 1.2 shows the approximate share each major company has of the global herbicide market. The top three companies have launched three very successful compounds that have each managed to establish and maintain a large share of a huge international market – Ciba with atrazine, Monsanto with glyphosate and Zeneca (previously ICI) with paraquat.

Table 1.2 *Estimated share of world herbicide market*

% Share of world market	Companies
> 10	Ciba, Monsanto, Zeneca
5–10	AgrEvo, BASF, Cyanamid, DowElanco, Du Pont, Rhone-Poulenc
2–5	Bayer, Kumiai, Sandoz

The fungicide market worldwide has always been of lower value than either herbicides or insecticides. It is often said that this is because farmers cannot see the causal agents of the diseases that damage their crops and so they do not treat for them. Whilst this may be true of developing nations where the education of farmers may be lower than that of the developed world, this cannot be true of the USA, and in North America fungicide use is very low indeed. In these extensive agricultural systems, it is usually factors such as lack of water rather than attack by disease that reduces yield. It is also true that it was not until the last thirty years that compounds became available to demonstrate the catastrophic losses that can be associated with disease. If you cannot grow a crop in the absence of disease how can you show the benefits of disease control? Couple this with the use of conventional disease resistance breeding and it becomes clear that fungicides are the new boys in crop protection, and as we learn more about the impact of poor disease management on crop productivity we understand the value of disease control. Once again, it is the companies that discovered the lead compounds that have the largest share of the fungicide market. Ciba with propiconazole (licensed from Janssen) for use in cereals and metalaxyl for control of Oomycete pathogens and Bayer with triadimefon, the first really successful, eradicant fungicide for use in cereals and other crops (Table 1.3).

Table 1.3 *Estimated share of world fungicide market 1995*

% Share of world market	Companies
> 15	Ciba
10–15	Bayer
5–10	AgrEvo, BASF, Du Pont, Rohm and Haas, Sandoz, Zeneca
2–5	Cyanamid, DowElanco, Hokko, Kumiai, Nippon Soda, Rhone-Poulenc, Sankyo, Sumitomo

The global insecticide market is a key market in the developing world and in high value crops such as fruit and vegetables and in cotton where damage to the flower or developing boll, the square, can lead to a complete loss of yield. Certain crops can be expected to suffer from insect attack on a routine basis. Maize planted in the mid-western United States will succumb to corn rootworm (*Diabrotica* spp.) unless the soil is treated, and Colorado potato beetle (*Leptinotarsa decimlineata*) will attack potatoes in North America and Continental Europe. The key company in the development of insecticides is Bayer, because of its strength in the area of organophosphorus chemistry (Table 1.4).

Table 1.4 *Estimated share of world insecticide market*

% Share of world market	Companies
> 10	Bayer
5–10	AgrEvo, Ciba, Cyanamid, DowElanco, FMC, Rhone-Poulenc, Zeneca
2–5	Sandoz, Sumitomo

Other products represent only a small share of the crop protection market with plant growth regulators being the largest of this sector and other product types including rodenticides, molluscicides, avicides and nematicides.

2.2 1995 Sales by Crop

The main crop protection markets include those high acreage crops that represent the bulk of the processed or fresh produce consumed by the world's population or its livestock, including oil crops. An exception to this is the cotton market that is important for the production of fibre for clothing, oil for food processing and protein for animal feed and is susceptible to insect attack. Figure 1.2 shows how the agrochemical sales are split between crops. The grain crops maize, rice, wheat and barley represent almost one-third of all chemical inputs, totalling more than the whole of the vegetable market. This is a representation of the size of the cultivated area of these crops and the high quality control standard that is applied to vegetable crops in today's world. When was the last time you found a caterpillar in a cabbage at the supermarket?

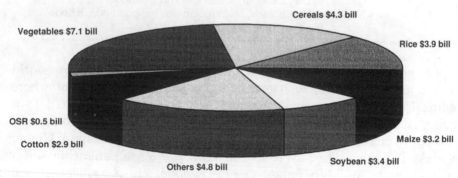

Figure 1.2 *World agrochemical market – sales by crop 1995*

One consequence of this split of agrochemical usage by crop is that agrochemical companies will always target established markets where they can expect to get a return from their investment. If you have invested $100 million on a new chemical and you expect its effective life in the market place to be ten years and you wish to make enough profit from that new product to ensure the security of your operation and your employees you will need to make a profit of over $10 million each year. This will mean sales of over $25 million each year. If a new compound achieves 10% of the sales in a particular field it is doing very well, so any market that is less than $250 to $500 million is too small for investment. Hence the large markets remain as large markets with intense competition and increasingly high standards of biological effect and low environmental impact.

2.3 1995 Sales by Region

Today the Eastern European market remains depressed by economic considerations and represents less than 3% of the total global agrochemical sales. Although the Eastern bloc has significant potential in the crop protection industry it is no longer recorded as a separate region rather contributing to the rest of the world (RoW) figure. The North American share of the market continues to grow and Western Europe shows an increased market share in dollar terms but much of this increase is due to exchange rate factors. The Far East's share of the market has also increased in dollar terms but again much of this was due to currency fluctuations. The Japanese market continues to contract due, in part, to currency fluctuations and to reduced rice plantings. Increased herbicide usage in Australasia and Latin America contributed to an increase in these regions' share of the agrochemical market.

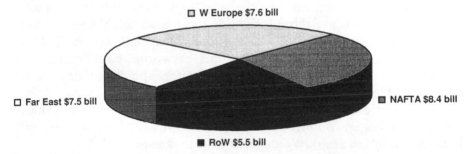

□ W Europe $7.6 bill

□ Far East $7.5 bill   ■ NAFTA $8.4 bill

■ RoW $5.5 bill

Figure 1.3 *1995 Agrochemical sales by region*

As with the review of agrochemical usage on crops, it is the developed world that consumes the majority of the pesticides produced. As these countries are the richest nations it is unlikely that this will change within the foreseeable future. This again means that agrochemical companies will target established markets in wealthy countries, but with an eye on the situation in large potential markets such as China (population 1 200 million) and India (population 900 million).

3 BIOLOGICAL SCREENING – DISCOVERY AND DEVELOPMENT OF A NEW AGROCHEMICAL

So far, it has been established that there is a very large market for crop protection agents globally and that these markets are dominated with crops such as wheat, barley, maize, soybeans, rice, cotton, grapes, top fruit and total weed control in countries and regions such as the USA,

Western Europe and Japan. The opportunities in South and Central America are beginning to be realised and, as the economics of Eastern Europe and the former Soviet Bloc improve, there is considerable potential in these areas as well. Countries such as India and China will also become significant opportunities for the development of agrochemicals but the restraints on trade that include such things as non-conformity with international patent law (although the introduction of GATT legislation – General Agreement on Tariffs and Trade – and acceptance into the league of friendly trading nations is changing this) has restricted penetration into these markets. For more details of the approaches to screening and early field testing see Copping *et al.*[11] and Copping.[12]

So how do companies search for new agrochemicals? In simple terms, each company has a list of key crop areas and a list of insects, pathogens and weeds that infest those crops. The targeted crops are usually those that command significant agrochemical input and the target pests, diseases and weeds are those for which there is an established market. For example, insects attack cotton (bollworms, boll weevils, aphids, thrips, jassids, whitefly), maize (cutworms, earworms, corn borers, corn rootworm), rice (leafhoppers), top fruit (mites, aphids, codling moth) and vegetables (caterpillars, aphids, whitefly), fungal pathogens infest small grain cereals (powdery mildew, rusts, *Septoria*, eyespot), rice (blast and blight), grapes (*Botrytis*, powdery and downy mildew) and top fruit (powdery mildew, scab). Weeds, of course, invade all fields laid down to the monoculture of a single crop as such single crop culturing is unstable in environmental terms and will invariably require inputs of energy to maintain. This energy input can be in the form of mechanical energy (from cultivation), manual energy (hand-weeding) or chemical energy (herbicides).

There are several ways of finding new compounds to test in an agrochemical screen and this chapter will not address these in any detail. It is sufficient to say that compounds may be prepared along an identified area of chemistry by synthetic chemists employed by the agrochemical company (with a view to identifying new chemical classes with novel biological spectra that can be patented), from the random selection of compounds from sources outside the company (usually bought in from Universities, specialist chemical companies or from chemical companies not involved in agrochemical research) or from natural sources (higher plants, fungi, bacteria, algae, marine organisms).

Biological screening can be divided into three different levels, and each compound will be tested at one of these levels, dependent upon what is known about it. If the compound has no known relationship with any previously tested compound that has shown activity in earlier tests it will enter at Phase 1. Phase 1 testing is designed to reject as quickly as possible all inactive compounds, so that little effort is spent on the test. Such tests are high dose and high volume against selected 'indicator species'. Indicator species are those that have been selected either because they represent a significant market in their own right or because they represent a group of pests, diseases or weeds that make up a potentially significant agrochemical market. Typical Phase 1 test organisms might be:

Herbicides	Barnyard grass (*Echinochloa crus-galli* – a non-temperate grass);
	wild oats (*Avena fatua* – a temperate grass);
	chickweed (*Stellaria media* – a temperate small seeded broad-leaf);
	mayweed (*Tripleurospermum maritimum* – a temperate large seeded broad-leaf);
	purslane (*Portulaca oleracea* – a non-temperate small seeded broad-leaf);
	morning glory (*Ipomoea purpurea* – a non-temperate large seeded broad-leaf);
	nutsedge (*Cyperus esculentus* – a non-temperate sedge).
Fungicides	Vine downy mildew (*Plasmopara viticola*);
	potato late blight (*Phytophthora infestans*);
	wheat (or barley) powdery mildew (*Erysiphe graminis*);
	grey mould (*Botrytis cinerea*);
	wheat blotch (*Septoria tritici*);
	rice blast (*Pyricularia oryzae*);
	striped rust (*Puccinia striiformis*);
	rice sheath blight (*Rhizoctonia solani*).
Insecticides	Vetch aphid (*Megoura viciae*)
	bollworm (*Helicoverpa zea*);
	army worm (*Spodoptera littoralis*);
	diamondback moth (*Plutella xylostella*);
	mustard beetle (*Phaedon cochleariae*);
	corn rootworm (*Diabrotica undecimpunctata*);
	whitefly (*Bemisia tabaci*);
	red spider mite (*Tetranychus urticae*).

The selection of the species used in a primary screening programme will depend upon the facilities available to the organisation, the local or regional restrictions on the cultivation or breeding of pathogenic species or phytophagous insects and mites and on the interests of the company. For example, if a company has a significant presence commercially in a particular crop area with existing products, it is likely that this crop and its pests and diseases will feature more strongly than they would if the company were poorly represented in that crop. The reason for this is because the company already has a major presence in that crop, the introduction of a new product will be easier than entering a completely new crop where it has no representation. It is always good to reinforce the commercial products available where you already hold a significant market share.

If a compound is not found to be inactive in the primary screen (or if it comes from an area of chemistry that has been found to show consistent biological activity) then it will enter a secondary screen. Here the questions being asked are different to those of the primary screen. Instead of 'is this compound inactive?' the question is 'how active is this compound against the target organisms in my screen and how does it compare with standard compounds?' The standard compounds may be typical commercial products that have a proven level of effect or selected best compounds from the chemical group(s) under evaluation, an 'internal standard', or both. The argument for the commercial standard is that it confirms the validity of the test and allows comparisons to be made from month to month between compounds screened at different times of the year. The argument for the best internal standard is that the commercial potential for the best compound tested to date is known and it is hoped that further synthesis and testing will improve biological activity, crop selectivity, environmental impact or all of these.

Secondary screening will look at reduced rates in lower volume sprays and with greater replication. Relative activity against target organisms (LC_{50} figures) can be determined against some organisms. If the biochemical mode of action is known, then the biological effect can be compared with the activity against the target enzyme. All these data will be compared with physico-chemical characteristics such as volatility, partition coefficient, solubility (in water and organic solvents), photostability and soil half-life. Many of these parameters will not be determined for each potentially active compound but they can be calculated within the accuracy of an order of magnitude, thereby giving an idea of the probable behaviour in the field. These tests enable compounds with preferred biological activities or cost advantages or reduced environmental risk to be selected for further study.

The further study is whatever you want it to be bearing in mind that the tertiary screen is designed to allow sound decisions to be made on the selection of compounds for transition into field development programmes. These development programmes are costly and the correct choices have to be made. Hence, tests consist of looking at formulations in comparison with earlier compounds, looking at crop selectivity, timing of application, systemicity, spectrum of effect, lowest dose that has an effect, soil persistence, effects on non-target crops and beneficial organisms and so on. They must be designed to ensure that only compounds with an advantage are taken further although there are arguments for testing a compound that falls below the expected level of activity if it is the first within a chemical class.

The synthesis, screening and field evaluation of compounds is a complex, expensive, time-consuming but exciting activity. There are many conference volumes written on this subject and the reader is directed to Makepiece *et al.*[13], Copping *et al.*[14] and Hewitt *et al.*[15]

4 QUESTIONS

1.1 In how many ways does crop protection improve the agricultural productivity of the world? What are these ways? (Hint, you should have at least five.)

1.2 If you were the research director of a large multinational company on which crops would you ask your research staff to concentrate, in which countries and why? (Hint, you should have at least five crops.)

1.3 If you ran the United Nations Food and Agriculture Organisation, dedicated to increasing productivity in developing countries, on which crops would you expect your staff to be working and in which countries?

1.4 On which single crop is the most money spent for crop protection?

1.5 What area of crop protection is the largest?

1.6 Why are photostability and volatility important properties for an agrochemical?

5 REFERENCES

1. D. Pimentel, L. McLaughlin, A. Zepp, B.L. Kitan, T. Kraus, P. Kleinman, F. Vancini, W.J. Roach, E. Graap, W.S. Keeton and G. Selig, 'Environmental and Economic Impacts of Reducing US Agricultural Pesticide Use', in D. Pimentel (ed.), 'Handbook of Pest Management', Vol. 1, 2nd. Edn., CRC Press, Boca Raton, 1991, pp. 679–719.

2. H.M. LeBaron, 'Weed Science in the 1990s: Will it be Forward or in Reverse?' *Weed Technol.*, **4**, 1990, 671–689.

3. N.E. Bourlag, 'The Challenge of Feeding 8 Billion People', *Farm Chem. Int.*, 1990, Summer Issue, pp. 10–12.

4. World Health Organisation, 'Public Health Impact of Pesticides Used in Agriculture', WHO Geneva, 1990.

5. National Research Council Board on Agriculture, 'Alternative Agriculture', Committee on the Role of Alternative Farming Methods in Modern Production Agriculture, Washington D.C., National Academic Press, Washington, DC, 1989.

6. E.G. Smith, R.D. Knutson, C.R. Taylor and J.B. Penson, 'Impacts of Chemical Use Reduction on Crop Yields and Costs', College Station, Texas, Agricultural and Food Policy Center, Department of Agricultural Economics, Texas A & M University System, 1990.

7. R.D. Sweet, J.E. Dewey, D.J. Lisk, W.R. Mullison, D.A. Rutz and W.G. Smith, 'Pesticides and Safety of Fruits and Vegetables', Comments from CAST, Ames, Iowa, Council for Agricultural Science and Technology, **1990–91**, 1990.

8. F. Urban and A.J. Dommen, 'The World Food Situation in Perspective', in 'World Agriculture Situation and Outlook Report', Washington D.C., National Agrochemicals Association, 1989.

9. F. Urban, 'Agricultural Resources Availability', in 'World Agricultural Situation and Outlook Report', Washington D.C., US Department of Agriculture, **WAS-55**, 1989, pp. 8–16.

10. 'Agrochemicals Executive Review', Allan Woodburn Associates Ltd, 18 Newmills Crescent, Balerno, Edinburgh, 1996.

11. L.G. Copping, H.G. Hewitt and R.R. Rowe, 'Evaluation of a New Herbicide', in 'Weed Control Handbook: Principles (8th Edn.)', ed. R.J. Hance and K. Holly, British Crop Protection Council, Farnham, UK, 1990, pp. 261–300.

12. L.G. Copping, 'Aspects of Pesticide Discovery', in 'Recent Developments in the Field of Pesticides and their Application to Pest Control', ed. K. Holly, L.G. Copping and G.T. Brooks, United Nations Industrial Development Organisation, Vienna, 1990, pp. 16–26.

13. R.J. Makepiece, J.C. Caseley and L.G. Copping (eds.), 'Influence of Environmental Factors on Herbicide Performance and Crop and Weed Biology', *Aspects of Applied Biology*, **4**, Association of Applied Biologists, Wellesbourne, UK, 1983.

14. L.G. Copping, C.R. Merritt, B.T. Grayson, S.B. Wakerley and R.C. Reay, (eds.), 'Comparing Laboratory and Field Pesticide Performance', *Aspects of Applied Biology*, **21**, Association of Applied Biologists, Wellesbourne, UK, 1989.

15. H.G. Hewitt, J.C. Caseley, L.G. Copping, B.T. Grayson and D. Tyson (eds.), 'Comparing Laboratory and Field Pesticide Performance II', British Crop Protection Council Monograph 59, BCPC, Farnham, UK, 1994.

CHAPTER 2

Herbicides

1 INTRODUCTION

Losses due to weed competition have been known since mankind first gave up a hunter/gatherer existence for that of a farmer. The concept of a piece of ground that is dedicated to the cultivation of a single plant species is one that flies in the face of Nature. This is because monoculture is inherently unstable through the intense competition between identical species for nutrients, water and light and because evolution demands competition for the available resource with those species best suited to the prevailing conditions dominating that particular environment. Man has selected his food crops from the many thousand plant species that exist for their nutritional and flavour characteristics rather than through their ability to compete. This means that in a monoculture crop situation there has to be an input of energy to ensure that the planted crop survives. Traditionally it has been relatively easy to remove weeds by hand as the crops were usually planted in rows and were uniform in appearance and in emergence. There was also an abundance of labour on the land to remove these weeds.

It was usual in agricultural systems of the 19th century to use crop rotations. These were based on the so-called Norfolk four-course rotation of root crops, barley, seeds and wheat. The root crop was introduced as it was a crop grown in wide rows that allowed the crop to be hoed – a cleaning crop – and the seed crop usually contained clover to replace nitrogen through the nitrogen fixing bacteria associated with the legume. Cereals were traditionally a dirty crop as they could not be hand weeded without significant damage to the crop.

As the industrial revolution drew workers from the fields to the cities, it became more difficult to rely on hand weeding to keep weeds at bay.

There are many reasons that weeds need to be controlled. The direct competition that leads to yield loss is the most obvious. Others include the contamination of the crop with seeds from aggressive or poisonous weeds and the effect of weed populations on the speed and efficiency of harvest. When seed is sold it is checked for contaminants and the value of the crop, or indeed whether it can be sold at all, is determined by the absence of foreign seeds. Wild oat seed (*Avena fatua*) is a seed that is serious problem in a cereal crop. If the farmer uses seed containing wild oat seed to sow the next year's crop, it is possible that he will be introducing a new weed to an area that had been free from it. It also reduces the value of the seed as feed and for processing. There are stories, that may be anecdotal, that speak of sorghum seed being contaminated by seed from the thorn apple (*Datura stramonium*). The trouble here is that the seeds are the same size and the same colour and so it becomes very difficult to separate them. Thorn apple seeds are very poisonous and their presence in a seed crop used for bread making is clearly undesirable and has been reported as being fatal. It is also common to grow peas for freezing in such a way that they are harvested, frozen and packed in a single operation within three hours. The flower head of the mayweed plant (*Matricaria matricarioides*) is exactly the same size as a mature pea and the processing machine is unable to separate the flower heads from the crop. The same is true for the fruit of black nightshade (*Solanum nigrum*).

Many weeds are very vigorous in their habit, often overtopping the crop in which they are growing. Such a weed is cleavers (*Galium aparine*) and it is common in many crops, particularly cereals. It is not unusual in a heavily infested field to see the weed growing above the crop and causing the crop to lodge or collapse under the weight of the weed. When the crop is ripe and ready for combining, the weed is often still green, is always very dense and overtops the cereal. This means that the infested crop has to be cut slowly with the cutter blade as close to the ground as possible in an attempt to collect all the seed. The weed will also wrap around the moving parts of the harvester and will require removing periodically so that the harvesting operation can continue. This is a slow process and often increases the time taken to harvest a crop by a factor of three or four. The more time taken to harvest the crop the more expensive that process becomes.

With the loss of labour and the increased intensity of farming, it became increasingly difficult to control weeds on farms and so the search began for chemicals that could be used selectively. The first examples included compounds such as copper sulfate (introduced in France in 1896 and in Britain in 1898) and were used to control weeds in cereals.[1] This was followed from 1901 to 1919 with compounds such as ferrous sulfate, sulfuric acid and sodium chlorate. The principle that applied in the case of these compounds was that the cereal was less easy to wet than the broad-leaved weeds growing therein and as such received less of the chemical and, therefore, survived while the weeds succumbed. Sulfuric acid is still used today, not as a selective herbicide but as a defoliant for crops such as potatoes, while sodium chlorate is used (in conjunction with a flame retardant) as a total herbicide and ferrous sulfate is often included in moss killers. The compounds were used selectively but they were not selective herbicides in their mode of action.

Dinitrophenols had been used as insecticides since 1892 but it was not until the 1930s that their value as herbicides was discovered and 4,6-dinitro-*o*-cresol (DNOC) was introduced. The trouble with dinitrophenols was their toxicity to all living organisms that respire. Their mode of action is through the uncoupling of oxidative phosphorylation, an effect that leads to a rapid death of any organism that comes into contact with the chemical, including the operator.

The revolution in selective weed control came about in the early 1940s with the discovery of the 'hormone herbicides' 2,4-D and MCPA (see Section 3.3 on Auxin-type herbicides). These compounds were truly selective and very active. They controlled broad-leaved weeds in cereal crops when applied post-emergence.

2 APPLICATION OF HERBICIDES

Today's herbicides can be applied in a variety of different ways to the crop. They can be applied as single component products or as combination products, using the different properties and weed spectrum of the components of the mixtures.

Post-emergence herbicides are applied after the emergence of the weed (and usually, but not necessarily, the crop as well). Compounds such as sulfuric acid must be applied such that they cover all the foliage of the target weed as they are contact herbicides. Others, such as the auxin-herbicides, are taken up by the weed and translocated throughout the target plant – they are systemic – and, consequently, it is not so important to ensure that the whole of the weed is covered. Some post-applied compounds are only active through the foliage, bentazone and the auxin-herbicides for example, whilst others are taken up through the roots following application, *e.g.* isoproturon. More recent compounds, such as the sulfonylureas, can be taken up through the foliage and the roots. Hence the fact that a herbicide is applied post-weed emergence does not indicate that the compound is taken up by the foliage it is merely a convenient description of the use of the compound. Post-emergence compounds can be applied to the entire crop/weed canopy. This is often described as an over-the-top application. Alternatively, the compounds can be directed away from the crop at the weeds – post-directed application.

Pre-emergence herbicides are applied pre-weed emergence and this will usually mean pre-crop emergence as well. Such compounds are taken up underground by the roots or hypocotyls of the weed. It is important for such compounds to possess some water solubility, in order that they become available to the germinating weed, but not so much that they are leached away from the weed germination zone. They must also be relatively persistent in the soil so that weeds that germinate over a period of time are all controlled.

Pre-plant incorporated herbicides are applied before the crop is sown and are incorporated into the soil. Hence, they are also applied before weeds emerge. The reason for incorporation is usually because the herbicides are volatile and would be lost if they were not incorporated, or light unstable and they would be degraded if they remained on the soil surface. Volatility is a useful characteristic as it allows the redistribution of the compound throughout the soil following incorporation.

3 MODES OF ACTION

If a biochemical process is peculiar to plants it is probable that this will be a useful target for a herbicide as its effect on non-target species will be minimal. The early compounds, such as copper sulfate, sulfuric acid and dinitrophenols, were effective because of the differential uptake between cereal crops and broad-leaved weeds. However, they also had toxic effects on other organisms that they contacted. Hence, compounds that are general toxicants or inhibitors of respiration through the uncoupling of oxidative phosphorylation, as a general rule, are not good herbicide candidates. (A good reference book on plant biochemistry, growth and function for the beginner is the Open University text by Irene Ridge.[2])

3.1 Photosynthesis

There are several processes that are, for the purposes of this chapter, exclusive to plants. One such process is photosynthesis.

It has been well-known for many thousands of years that life as we know it is dependent upon plants. "All flesh is grass, and the goodliness thereof is as the flower of the field" (Isaiah[3]). Plants, through photosynthesis, produce our food, our shelter, our clothing, our paper, the food for our animals and the oxygen that we breath.

Photosynthesis involves the conversion of light energy into chemical energy, the light reactions, and the incorporation of carbon dioxide into sugars, the dark reactions. The light reactions (Figure 2.1) capture light energy and convert this into chemical energy through the electron transport chain. The products of the light reactions are chemical energy in the form of ATP, reducing power in the form of NAPDH and oxygen as a by-product. The light reactions are divided into two cycles: photosystem I or cyclic photophosphorylation and photosystem II or non-cyclic photophosphorylation. Both involve the capture of light energy by chlorophyll, a photo-receptor and the acceptance of electrons from the splitting of water. The capture of these electrons increases the energy level of the chlorophyll to the so-called singlet state and this then returns to the ground state as the electrons flow through an electron transport chain to produce ATP and NADPH. If the electron transport chain is interrupted and light continues to fall on the chloroplast, the energy level of the chlorophyll is raised from the singlet state to the triplet state. Triplet chlorophyll can interact in a damaging way with membrane lipids but, more importantly, it can excite oxygen, there in abundance because of active photosynthesis, to a singlet state. This singlet oxygen is very reactive and it interacts with cellular lipids, proteins, nucleic acids and many other plant cell components thereby inducing cellular disorganisation and plant death. Some of the early herbicides inhibited non-cyclic photophosphorylation and it was believed that treated plants died because they did not photosynthesise and therefore starved to death. It

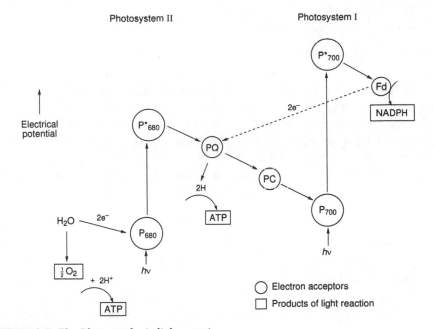

Figure 2.1 *The Photosynthetic light reactions*

was shown, however, that if a plant was treated with a photosystem II inhibitor and left in the light it died more rapidly than an untreated plant left in total darkness. If the treated plant was left in the dark, however, it died at the same rate as the untreated. There was clearly more to the plant's demise than starvation. It was later shown that it was the interference with the highly energetic light reactions that resulted in plant death.

A number of herbicides interfere with the flow of electrons from singlet chlorophyll through the electron transport chain. The main herbicide groups and examples from these groups are shown in Table 2.1.

Cyclic photophosphorylation is also a highly energetic reaction. The bipyridyliums, paraquat and diquat (Figure 2.2), divert the electron flow of cyclic photophosphorylation (photosystem I). The capture of an electron from the chlorophyll reduces the herbicide and the reduced herbicide reacts with oxygen to form superoxide. Superoxide produces hydrogen peroxide within the chloroplast and these two compounds interact to form hydroxyl radicals in the presence of an iron catalyst. Hydroxyl radicals are very damaging and lead to the destruction of the cellular components leading to rapid plant death.

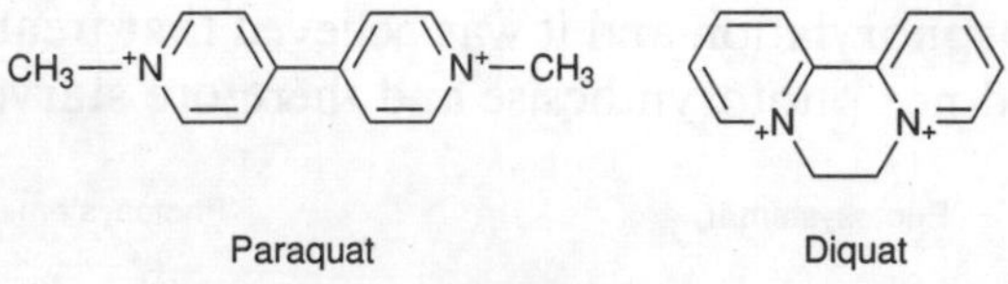

Paraquat   Diquat

Figure 2.2 *Paraquat and diquat*

The dark reaction of photosynthesis (Figure 2.3) is so-called as it does not require light to proceed. It does, however, require the products of the light reaction to operate and it will not, therefore, take place in the absence of light. It was discovered by Calvin and is often known as the Calvin cycle.

The dark reaction involves the fixation of carbon dioxide to form carbohydrates. The ATP and the NADPH produced in the light reaction drive this carbon fixation. It might be thought that the interruption of the Calvin cycle would also produce effective herbicides but this is not the case. There are two reasons why. First, the reaction is not an energetic reaction whose interruption would lead to the destruction of cellular components and second, the enzymes involved in the process are present in very high amounts. If an enzyme is to be targeted as a key step in the metabolism of a plant, it is important that it is present in small amounts and that it is not turned over very quickly. If an enzyme is abundant,

Table 2.1 *Herbicides that interupt electron flow in non-cyclic photophosphorylation (photosystem II)*

Herbicide class	Example	Date of introduction		
1,3,5-Triazine	Atrazine	1958	$R^1=CH_2CH_3$	$R^2=CH(CH_3)_2$
	Simazine	1956	$R^1=CH_2CH_3$	$R^2=CH_2CH_3$
	Cyanazine	1970	$R^1=CH_2CH_3$	$R^2=$ 2-methylpropio-nitrile
	Terbuthylazine and many more	1967	$R^1=CH_2CH_3$	$R^2=C(CH_3)_3$

$$R^1-\underset{H}{N}\diagdown\diagup N-R^2 \text{ (triazine ring with Cl)}$$

Phenylurea			R^1	R^2	X
Phenylurea	Diuron	1959	CH_3	CH_3	3,4-Cl_2-Phenyl
	Linuron	1962	CH_3	OCH_3	3,4-Cl_2-Phenyl
	Chlorotoluron	1971	CH_3	CH_3	3 Cl,4 CH_3-Phenyl
	Fluometuron	1965	CH_3	CH_3	3-CF_3-Phenyl
	Isoproturon	1975	CH_3	CH_3	4-$CH(CH_3)_2$-Phenyl
	Methabenzthiazuron and many more	1968	CH_3	H	Benzothiazole

$$X-\underset{H}{N}-\overset{O}{\overset{\|}{C}}-N\diagup^{R^1}_{R^2}$$

Acylanilide	Propanil	1961

(structure: 3,4-dichlorophenyl-$NHCOCH_2CH_3$)

			X
Hydroxy-benzonitrile	Ioxynil	1970	I
	Bromoxynil	1971	Br

(structure: phenol with CN, and X at 3,5 positions, OH)

			R	X
Phenyl carbamate	Phenmedipham	1968	CH_3	CH_3
	Desmedipham	1970	CH_2CH_3	H

(structure: X-phenyl-$NHCO_2$-phenyl-$NHCO_2R$)

(Continued)

Table 2.1 *continued*

Herbicide class	Example	Date of introduction		
Uracil	Bromacil	1965		
	Lenacil	1965		

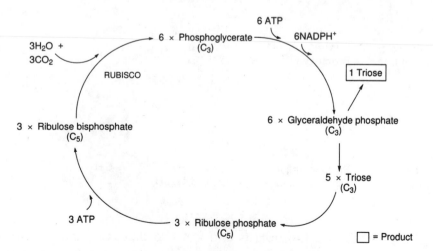

			R^1	R^2
Triazinone	Metribuzin	1972	$C(CH_3)_3$	SCH_3
	Metamitron	1977	Phenyl	CH_3

Miscellaneous	Bentazone	1972		
	Chloridazon	1962		

Figure 2.3 *The dark reaction (the Calvin cycle) of photosynthesis*

then a large amount of inhibitor will be needed to prevent its functioning and if it is being biosynthesised and degraded rapidly then, again, the effect of an inhibitor will be limited. The enzyme that combines carbon dioxide and ribulose 1,5-bisphosphate, ribulose bisphosphate carboxylase (RUBISCO), is claimed to be the most abundant protein on earth. Hence, it is not a good target for a herbicide.

There are a number of other herbicides that affect photosynthesis indirectly. Pyrazole herbicides such as benzofenap, pyrazolynate and pyrazoxyfen interfere with chlorophyll biosynthesis and have found commercial application for the control of annual and perennial weeds in paddy rice and maize (Figure 2.4).

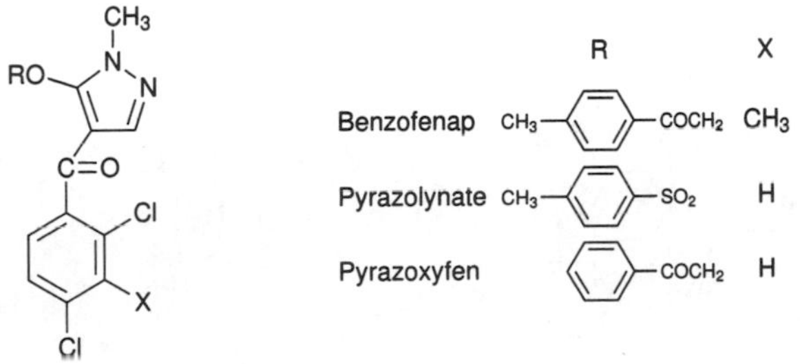

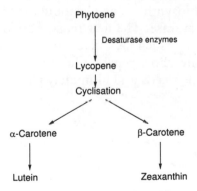

Figure 2.4 *Pyrazole herbicides*

In addition to the green chlorophyll pigments in the leaf's chloroplasts, there are other pigments that can also capture light energy but that also protect the leaf from damaging radicals by quenching them. Carotenoids are examples of this type of pigment (Figure 2.5).

Phytoene

↓ Desaturase enzymes

Lycopene

↓ Cyclisation

α-Carotene   β-Carotene

↓   ↓

Lutein   Zeaxanthin

Figure 2.5 *Part of the biosynthetic pathway to carotenoids*

The inhibition of carotenoid biosynthesis removes these protective pigments from the chloroplasts and leads to damaging effects within them. Herbicides that have been shown to interfere with carotenoid biosynthesis include norfluazon, fluridone and diflufenican (Figure 2.6). Herbicides that contain 3-trifluoromethylphenyl substituents (for example, fluometuron – Table 2.1) have also been shown to affect carotenoid biosynthesis. These compounds interfere with the desaturase enzymes that convert phytoene into lycopene whereas amitrole (Figure 2.6) prevents the cyclisation of lycopene to form the carotenes.

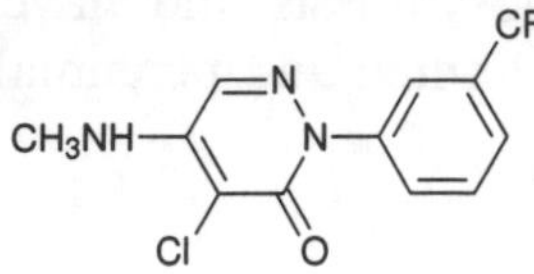

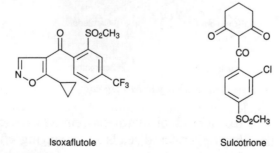

Norfluazon

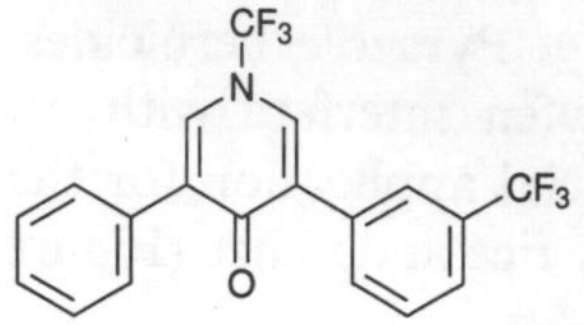

Fluridone

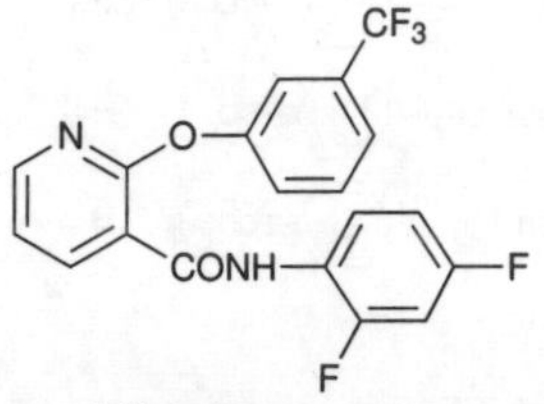

Diflufenican

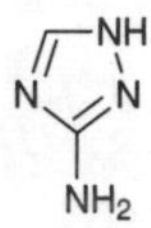

Amitrole

Figure 2.6 *Herbicides that interfere with carotenoid biosynthesis*

The enzyme *p*-hydroxyphenylpyruvate dioxygenase is involved in the conversion of *p*-hydroxyphenylpyruvate into homogentisate, a key step in plastoquinone biosynthesis. Inhibition of this enzyme has an indirect effect on carotenoid biosynthesis as plastoquinone is a co-factor of the enzyme phytoene desaturase. The new maize herbicide isoxaflutole and the triketone herbicides such as sulcotrione (Figure 2.7), inhibit *p*-hydroxyphenylpyruvate dioxygenase and this leads to the onset of bleaching in susceptible weeds and ultimately plant death.[4]

Isoxaflutole

Sulcotrione

Figure 2.7 *Isoxaflutole and sulcotrione*

There are several products that exert their effect through the accumulation of abnormally high levels of chlorophyll precursors.[5] A structurally diverse range of herbicides has been shown to inhibit the enzyme protoporphyrinogen oxidase, a pivotal enzyme at the branching point of the porphyrin pathway leading to both haeme and chlorophyll biosynthesis. The inhibitors of this process can be classified into three major chemical groups: the nitro diphenyl ethers (acifluorfen, lactofen), the phenyl heterocycles (oxadiazon and sulfentrazone) and the heterocyclic phenylimides (flumiclorac) (Figure 2.8).

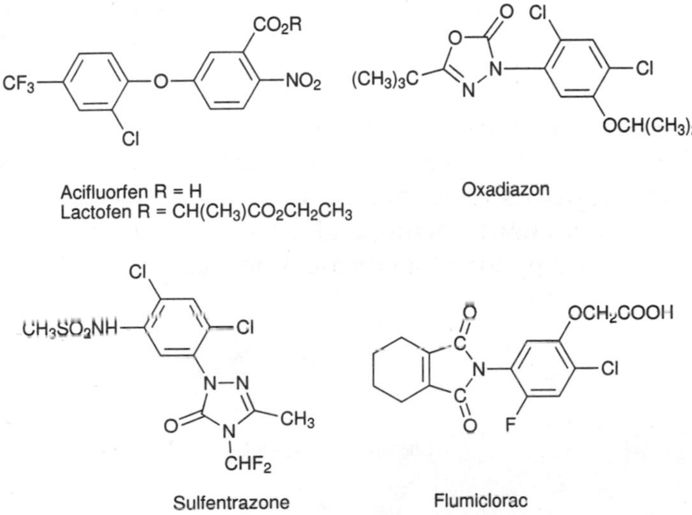

Acifluorfen R = H
Lactofen R = CH(CH₃)CO₂CH₂CH₃

Oxadiazon

Sulfentrazone

Flumiclorac

Figure 2.8 *Protoporphyrinogen oxidase inhibiting herbicides*

These compounds exert their effect through inhibition of membrane-bound chloroplastic protoporphyrinogen oxidase, leading to a transient accumulation of protoporphyrinogen IX. The protoporphyrinogen IX leaks out into the cytoplasm where it is converted into protoporphyrin IX by the herbicide-insensitive plasma membrane protoporphyrinogen oxidase. This protoporphyrin IX reaches very high levels in or near the plasma membrane and, being a photodynamic pigment, generates highly reactive oxygen radicals in the cytosol. The plasma membrane is, therefore, rapidly destroyed, leading to cell death.

This mode of action has been shown to be very effective at controlling weeds with rates as low as 1 g ha^{-1} leading to plant death for two good reasons. In the first place, there is little substrate competition with the herbicide because the substrate is lost to the cytoplasm when inhibition occurs and, second, because protoporphyrin IX will accumulate even when only a small proportion of the chloroplast protoporphyrinogen oxidase is inhibited.

3.2 Amino Acid Biosynthesis

Animals acquire many of their nutrients in a ready made form from the food that they eat. Plants, however, have to biosynthesise everything that they need for efficient growth. The first section of this chapter discussed photosynthesis, a fundamental biosynthetic process, but plants also synthesise other components that animals do not. These biosynthetic processes are good examples of potentially plant selective herbicidal targets. Amino acids are the building blocks of proteins and as such their biosynthesis is one such process.

3.2.1 Aromatic Amino Acid Biosynthesis. The shikimate pathway is the biosynthetic route to the aromatic amino acids tryptophan, tyrosine and phenylalanine as well as a large number of secondary metabolites such as flavonoids, anthocyanins, auxins and alkaloids. One enzyme in this pathway is 5-enolpyruvyl shikimate-3-phosphate synthase (EPSP synthase) (Figure 2.9).

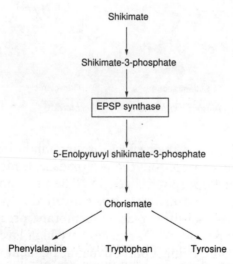

Figure 2.9 *Part of the biosynthetic pathway of aromatic amino acid synthesis*

The herbicide glyphosate (Figure 2.10) inhibits the enzyme 5-enolpyruvyl shikimate-3-phosphate synthase thereby preventing the biosynthesis of aromatic amino acids. This compound is used post-emergence and has been shown to be translocated in the plant's phloem to meristematic tissue, underground storage organs and stem apices. This property

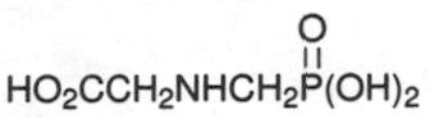

$$\text{HO}_2\text{CCH}_2\text{NHCH}_2\overset{\overset{\displaystyle O}{\|}}{\text{P}}(\text{OH})_2$$

Figure 2.10 *Glyphosate*

property allows the compound to control established perennial weeds in addition to annual weeds. It has no pre-emergence activity. A gene has been isolated from both a bacterium and a mutant petunia that codes for an enzyme that is not inhibited by the herbicide. Transgenic crops have been produced that contain this gene and are, thus, glyphosate-tolerant. Glyphosate is the only commercial herbicide with this mode of action.

3.2.2 Branched Chain Amino Acid Biosynthesis. The branched chain amino acids, leucine, isoleucine and valine, are produced by similar biosynthetic pathways (Figure 2.11). In one pathway, acetolactate is produced from pyruvate and in the other acetohydroxybutyrate is produced from threonine. Both reactions are catalysed by the same enzyme that is known as both acetolactate synthase (ALS) and acetohydroxy acid synthase (AHAS).

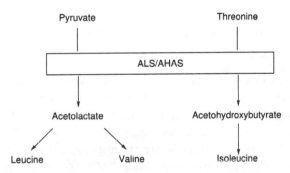

Figure 2.11 *Biosynthetic pathway to branched chain amino acids*

This enzyme is the target for a number of very active, low dose herbicides, including the sulfonylureas, triazolopyrimidine sulfonanilides, imidazolinones and pyrimidinyloxybenzoic acid analogues (Figure 2.12). Most of these compounds are applied post-emergence although some do have activity through the roots. As a general rule, however, the commercially available compounds today are insufficiently persistent in the soil to give season-long residual weed control.

3.2.3 Glutamine Synthetase. The enzyme glutamine synthetase is very important in the control of nitrogen metabolism in plants. It catalyses

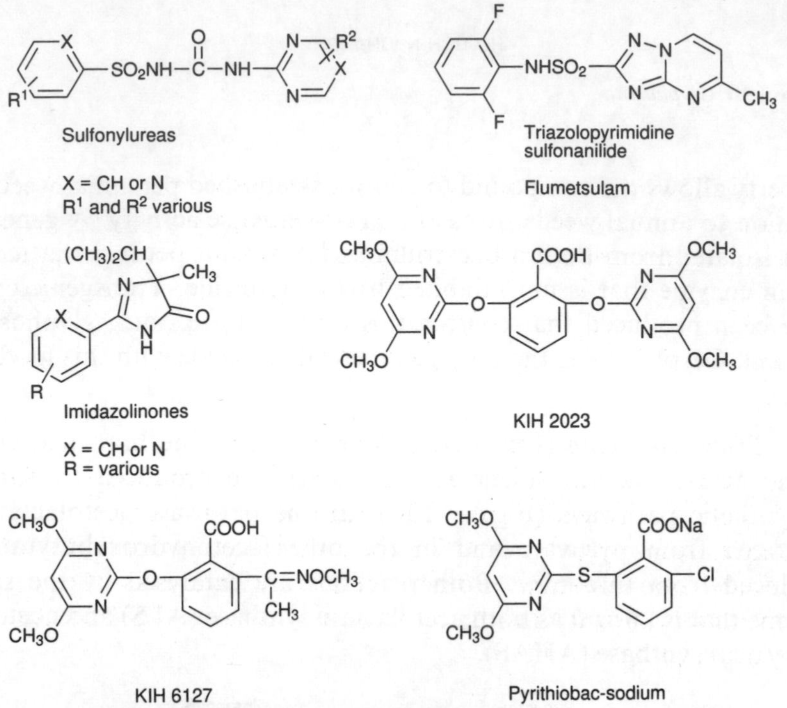

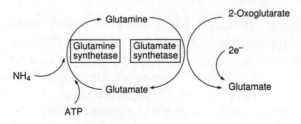

Figure 2.12 *Compounds that inhibit branched chain amino acid biosynthesis*

the combination of ammonia with glutamate to form glutamine. Gluta-mine is key in the transamination of keto-acids in the synthesis of several amino acids and is also an effective method of maintaining a low level of ammonia within the plant's cells (Figure 2.13). This enzyme is competi-tively inhibited by the transition state analogue glufosinate (Figure 2.14). This is a non-selective post-emergence herbicide that shows no pre-emergence activity as it is irreversibly bound to the soil.

Figure 2.13 *Biosynthesis of glutamine*

A close analogue of glufosinate, bilanafos, that is the L-homoalanyl-L-alanyl-L-alanyl amide of glufosinate, is produced by the soil strepto-

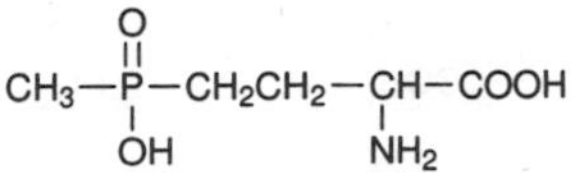

$$CH_3-\overset{\overset{\displaystyle O}{\|}}{\underset{\underset{\displaystyle OH}{|}}{P}}-CH_2CH_2-\underset{\underset{\displaystyle NH_2}{|}}{CH}-COOH$$

Figure 2.14 *Glufosinate*

mycete *Streptomyces hygroscopicus*. A gene has been isolated from this bacterium that acetylates the compound and thereby renders it non-phytotoxic. This gene has been used to transform crops and has been used as a selectable marker for the selection of crops transformed with genes coding for other traits and as a means of producing crops that are tolerant of over-the-top applications of glufosinate.

3.3 Auxin-type Herbicides

Compounds that control the growth and differentiation of plants are well-known and compounds that interfere with the function or that mimic the effects of such plant growth regulators would be expected to be effective as herbicides. Interference with the biosynthesis or expression of gibberellic acid would lead to a reduction of the height of weeds, rendering them less competitive to a crop not so retarded and would also provide good ground cover, thereby preventing the erosion of bare soil by wind and rain. However attractive such a policy might seem to be, no compound has been developed with this mode of action for use as a weed control agent. Many plant growth retardants discussed in Chapter 5 exhibit this mode of action though.

Indole-acetic acid is a plant growth regulator whose concentration in the plant is carefully regulated by synthesis, conjugation and degradation. It is believed that the auxin or hormone herbicides act by imitating the natural auxin but with no means of controlling the level of the synthetic auxin within the treated plant. Such compounds have been available to the farmer for over 40 years the first compounds being 2,4-D and MCPA, both discovered during the Second World War. They brought a revolution in weed control; being originally developed to control charlock (*Sinapis arvensis*) in cereals they showed good broad-spectrum broad-leaved weed control following post-emergence application and they were truly selective (unlike copper sulfate and sulfuric acid). A wide range of compounds with modes of action that are thought to be the same as the aryloxyalkanoic acids have been introduced since 1945 (Figure 2.15). Notable amongst these are the benzoic acids (dicamba) and the pyridinecarboxylic acids (clopyralid). Although the symptoms of all these compounds are similar, stem enlargement, callus

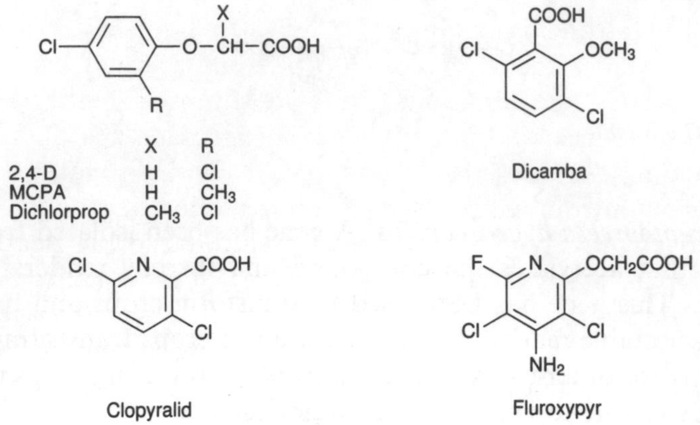

	X	R
2,4-D	H	Cl
MCPA	H	CH₃
Dichlorprop	CH₃	Cl

Dicamba

Clopyralid

Fluroxypyr

Figure 2.15 *Hormone or auxin herbicides*

growth, epinasty, leaf deformities and the formation of secondary roots, the absolute mode of action has yet to be confirmed. It is thought that the compounds act as auxins, binding to the auxin receptor in the sensitive weed, and that they continue to exert their effects because the plant is unable to reduce their concentration.

Recently, two new herbicides have been introduced that are quinoline carboxylic acid derivatives. Quinclorac and quinmerac (Figure 2.16) are effective through the formation of ethylene that is stimulated through the induction of 1-aminocyclopropane-1-carboxylic acid (ACC) synthesis, which leads to massive accumulation of abscisic acid. This results in reductions in stomatal aperture, water consumption, carbon dioxide uptake and plant growth.[6,7] Interestingly, quinclorac is effective at controlling barnyard grass (*Echinochloa crus-galli*) in rice culture while quinmerac controls weeds such as cleavers (*Galium aparine*) in a variety of different crops.

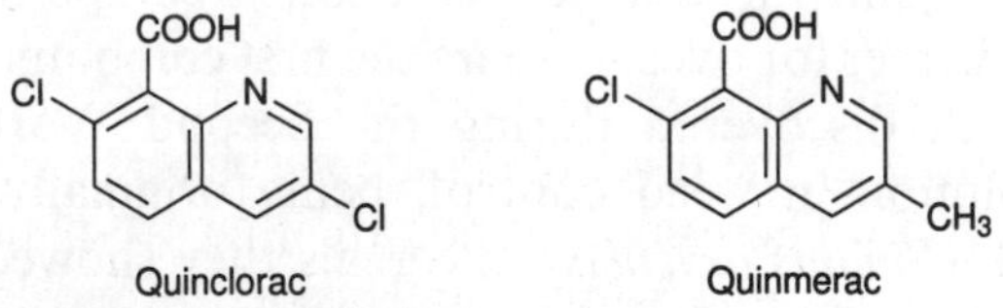

Quinclorac   Quinmerac

Figure 2.16 *Quinclorac and quinmerac*

3.4 Lipid Biosynthesis Inhibitors

Lipids are essential plant components as they are constituents of membranes and cuticular waxes as well as being major seed storage products. The fatty acid constituents of lipids are synthesised from acetyl coenzyme A under the influence of the enzyme acetyl coenzyme A carboxylase (ACCase) (Figure 2.17). Two groups of herbicide inhibit the action of ACCase, the aryloxyphenoxypropionates and the cyclohexanedione oximes (Figure 2.18). The failure to synthesise fatty acids and the subsequent membrane lipids leads to a cessation of growth, necrosis in the actively dividing meristematic tissue and plant death. It is interesting that these groups of compounds are very effective post-emergence treatments for the control of grass weeds in broad-leaved crops, although selectivity for some compounds in cereal crops has been introduced (see Section 4, Herbicide Selectivity). They have no activity against dicotyledonous or cyperaceous species.

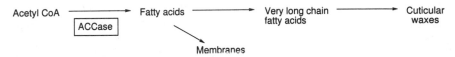

Acetyl CoA ⟶ Fatty acids ⟶ Very long chain fatty acids ⟶ Cuticular waxes

ACCase

Membranes

Figure 2.17 *Biosynthetic pathway to fatty acids, lipids and waxes*

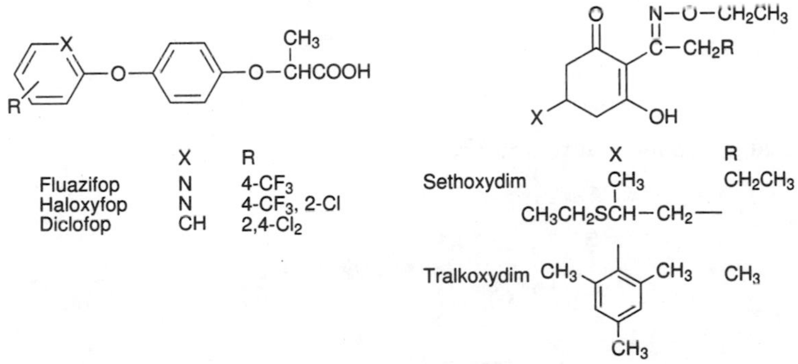

	X	R
Fluazifop	N	4-CF₃
Haloxyfop	N	4-CF₃, 2-Cl
Diclofop	CH	2,4-Cl₂

	X	R
Sethoxydim	CH₃	CH₂CH₃
	CH₃CH₂SCH—CH₂—	
Tralkoxydim	CH₃	CH₃

Figure 2.18 *Aryloxyphenoxypropionates and cyclohexanedione oximes*

The conversion of fatty acids into very long chain fatty acids is specifically inhibited by the thiocarbamate herbicides such as EPTC and triallate (Figure 2.19). These compounds are used pre-plant incorporated for the control of grass and some small seeded broad-leaved weeds in crops such as maize and small grain cereals.

[CH₃(CH₂)₂]₂NC—SCH₂CH₃

EPTC

[(CH₃)₂CH]₂NC—SCH₂CCl=CCl₂

Triallate

Figure 2.19 *Thiocarbamate herbicides*

3.5 Cell Division Inhibitors

Cell division is a fundamental prerequisite for plant growth. The meristematic regions of the plant are the targets of two major groups of herbicide that interfere with the organisation of the microtubules that are essential for the formation of the mitotic spindle along which the chromosomes separate during mitotic cell division. The microtubules are composed of both α-tubulin and β-tubulin that are brought together at the microtubule organisation centre to produce the microtubules themselves. The 2,6-dinitroanilines (Figure 2.20) interfere with the formation of the tubulins directly whilst the carbamates (Figure 2.21) prevent the organisation of the microtubule organisation centre itself. The result of this disruption is a failure of the cell division process and plant death. Both of these groups of compounds are only effective on germinating weed seeds and the majority are used pre-plant incorporated.

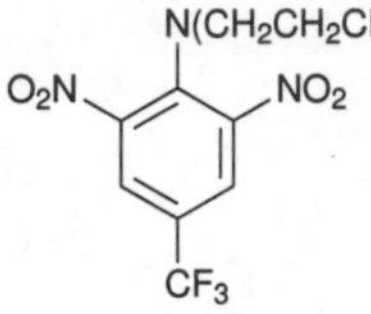

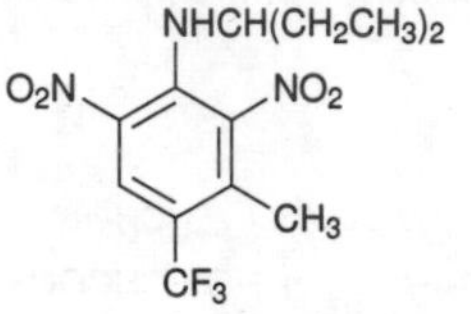

Figure 2.20 *2,6-Dinitroaniline herbicides*

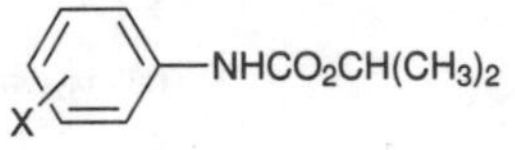

Figure 2.21 *Carbamate herbicides*

The 2-chloroacetanilides (Figure 2.22) are also suggested to inhibit cell division in susceptible weeds. These compounds have found a major commercial market for the pre-emergence control of grass and some small seeded broad-leaved weeds in crops such as maize and soybean. It is likely that 2-chloroacetanilides also alkylate the sulfhydryl groups of certain essential plant enzymes.

The rice herbicides cinmethylin (Figure 2.23), mefenacet (Figure 2.24) and daimuron and methyldymron (Figure 2.25) also interfere with meristematic activity in susceptible species.

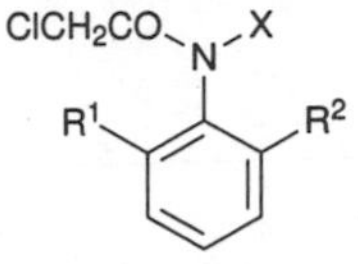

	R^1	R^2	X
Acetochlor	CH$_3$	CH$_2$CH$_3$	CH$_2$OCH$_2$CH$_3$
Alachlor	CH$_3$	CH$_3$	CH$_2$OCH$_3$
Metolachlor	CH$_3$	CH$_2$CH$_3$	CH(CH$_3$)CH$_2$OCH$_3$

Figure 2.22 *2-Chloroacetanilides*

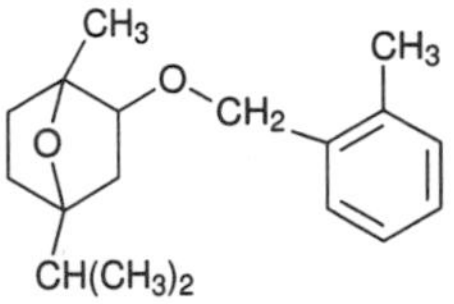

Figure 2.23 *Cinmethylin*

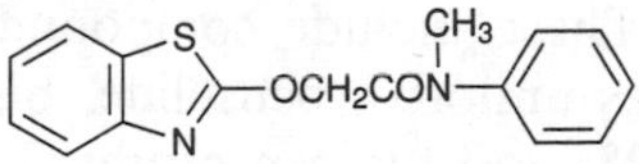

Figure 2.24 *Mefenacet*

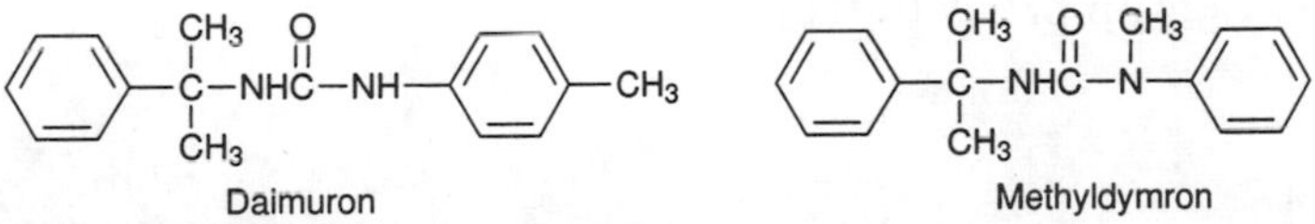

Daimuron

Methyldymron

Figure 2.25 *Daimuron and methyldymron*

3.6 Inhibitors of Cell Elongation

The *N*-arylalanine ester herbicides such as benzoylprop-ethyl and flamprop-methyl and difenzoquat (Figure 2.26) prevent cell elongation in certain grass weeds, allowing the crop to overtop them. The weeds are thus outcompeted and die. The exact mode of action is not certain but it is proposed that these compounds interfere with the site of action of the auxins.

3.7 Miscellaneous Modes of Action

The herbicide dichlobenil (Figure 2.27) is believed to exert its effect through the inhibition of cellulose biosynthesis of actively growing plant tissue, leading to a cessation of cell division and death.

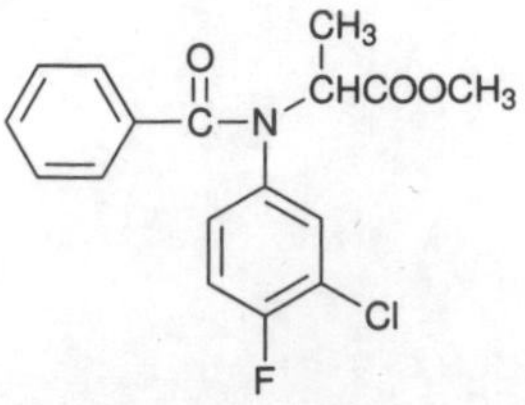

Figure 2.26　N-*Arylalanine ester herbicides and difenzoquat*

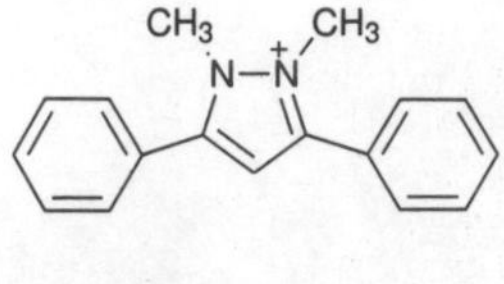

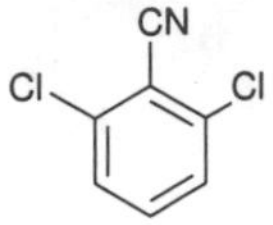

Figure 2.27　*Dichlobenil*

There are a number of herbicides whose mode of action is not known or that is uncertain. These include compounds such as the organo-phosphorous herbicides anilofos, bensulide, butamifos, fosamine and piperophos (Figure 2.28) and the benzofuranyl alkanesulfonates, etho-fumesate and benfuresate (Figure 2.29)

The Pesticide Manual[10] is a valuable source of information on the chemistry of crop protection agents.

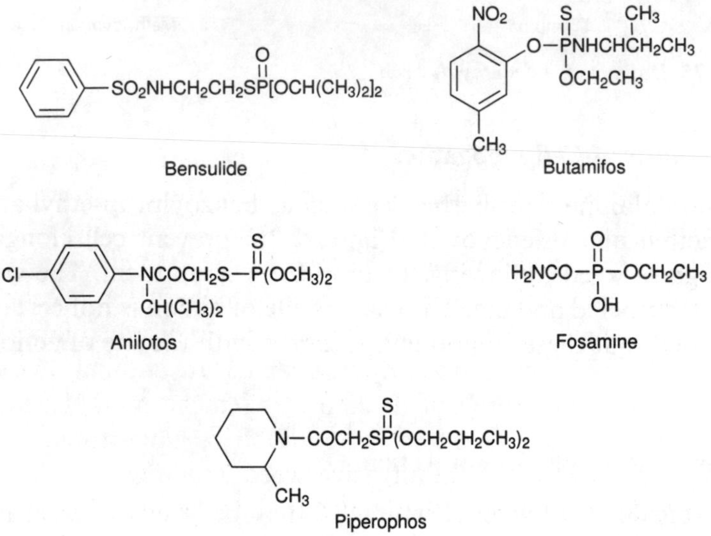

Bensulide

Butamifos

Anilofos

Fosamine

Piperophos

Figure 2.28　*Organophosphorus herbicides*

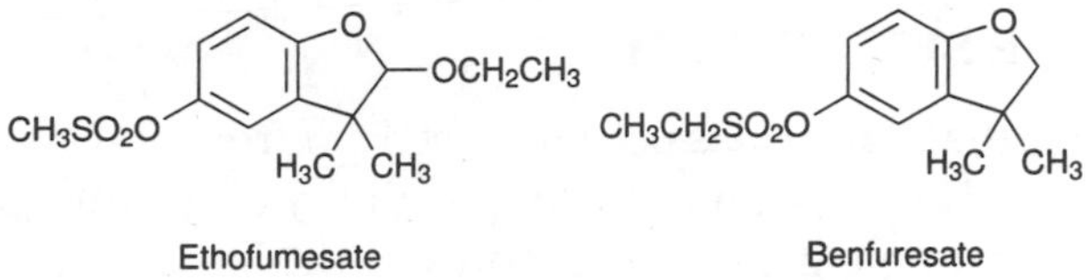

Figure 2.29 *Benzofuranyl alkanesulfonate herbicides*

3.8 Living Systems for Weed Control

It has been suggested that it would be environmentally friendly to reduce the input of chemicals into agriculture and replace them with natural systems that would control weeds by attacking them. Plant pathogens have been examined for this role and a number have been identified and commercialised. The problems associated with the introduction of such systems are enormous. It is important to confirm that the pathogen is specific for the targeted weed and that it will not invade the crop or related non-weed species. It must be aggressively pathogenic such that it will invade the host under all conditions, as the conditions that exist when the weeds need to be controlled are seldom ideal for fungal invasion. The formulated product must have a shelf-life that will enable the farmer to store it until it is needed and it should control the weed for the season in which it is applied and no longer, thereby persuading the farmer to purchase additional product the following season.

All plant pathogens must, necessarily, be weed specific and so the economics demand that hard-to-kill weeds in major or high value crops should be targeted. No product remains on the market today, but Collego (*Colletotrichum gloeosporoides* f. sp. *aeschynomene*) was sold for the control of northern joint vetch (*Aeschynomene virginica*) in soybeans and rice and Devine (*Phytophthora palmivora*) found a small market for the control of strangler vine (*Morrenia odorata*) in citrus orchards in Florida. For a review of mycoherbicides see Greaves.[8]

Other workers have examined the use of insects to consume troublesome weeds. Again there are problems associated with the selection of an insect that is specific to the weed and has no other disadvantages such as the sensitivity of some people. Such systems are unlikely to become widely used in agriculture but do find small niche markets, especially with the home gardener. Workers in South Africa have evaluated several insects for the control of the invasive weed *Lantana camara* and the aquatic fern *Salvinia molesta* with encouraging results.[9] The grass carp has also been recommended for control of a number of submerged aquatic weeds.

4 HERBICIDE SELECTIVITY

It is important that herbicides used for weed control in crops are selective to that crop. Early compounds used the inability of the foliage of upright cereal crops to retain a great deal of spray as the basis of their selectivity and this was successful in many situations although it must be remembered that the alternative to weed control with inadequately selective herbicides was hand weeding or no weed control at all.

There are a number of mechanisms of selectivity that are found in the herbicides that are used today. Diuron is used as a residual broad-spectrum herbicide in a number of situations such as plantations and forests. It is, however, phytotoxic to most perennial species and the selectivity shown by the established trees is because the compound does not move within the soil profile to a depth where established tree roots will absorb the compound in sufficiently high concentrations to exert an effect. This is selectivity by placement.

Other herbicides are selectively inactivated by the target crop whilst the weeds that they control either do not metabolise them or they do it so slowly that the weed is killed before it can inactivate the herbicide. There are a number of key plant enzymes that are used in the inactivation of herbicides. Microsomal mixed function oxidases are able to hydroxylate a wide range of herbicides such as bentazone and diclofop-methyl (Figure 2.30). It is often the case that these hydroxylated metabolites are subsequently glucosylated by sugars in the tissue and these conjugants can be stored in the cell vacuole where they can have no phytotoxic effects.

Deamination is another detoxifying metabolic process. Herbicides

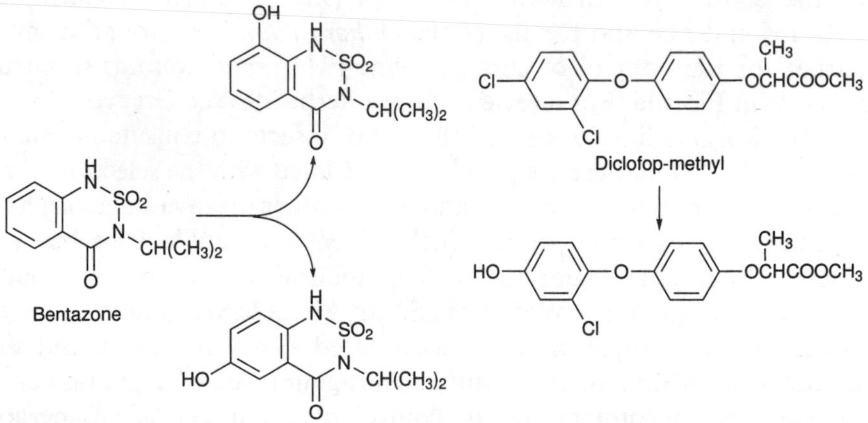

Figure 2.30 *Hydroxylation reactions*

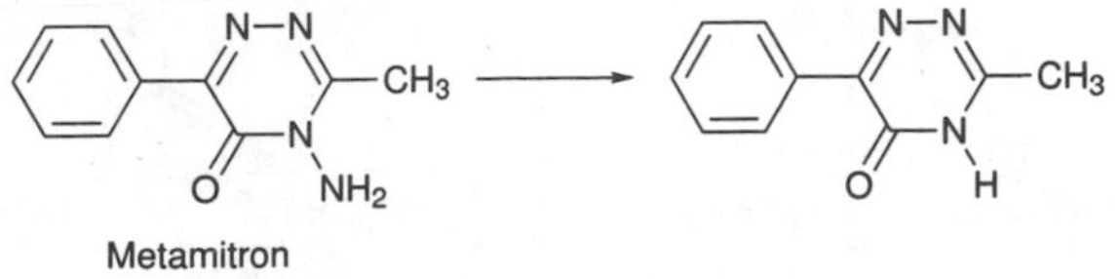

Figure 2.31 *Deamination of metamitron*

such as metribuzin and metamitron are deaminated almost certainly by a peroxisome-based deaminase to inactive intermediates (Figure 2.31).

Dealkylation is another common inactivating process. The phenyl urea herbicides and the 1,3,5-triazines are dealkylated to inactive metabolites almost certainly through the action of the microsomal mixed function oxidases discussed above (Figure 2.32). The 1,3,5-triazines also conjugate with glutathione to form inactive conjugates and a similar process is seen with thiocarbamates although the conjugation of these compounds demands the formation of the sulfoxide (Figure 2.33). All plants are able to perform this conjugation, although some use homoglutathione instead, and it is the speed with which the conjugation takes place that determines the damage to the plant. Some compounds, that themselves have no observable biological effect, have been shown to increase the levels of glutathione *S*-transferase in treated crops. The application of such compounds in conjunction with a herbicide will reduce the damage from the herbicide to the crop. Such safeners are finding increasing use in product formulations.

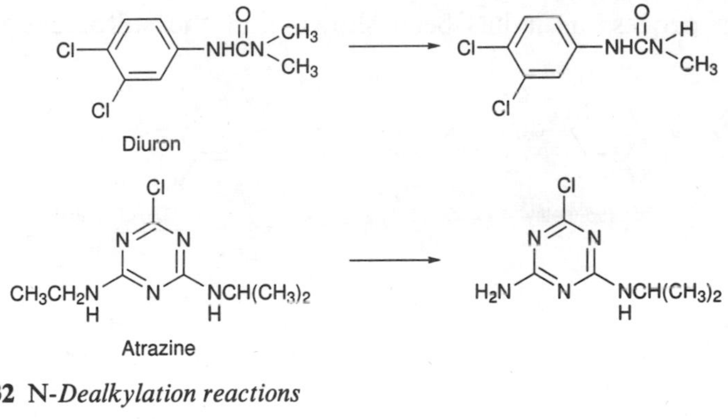

Figure 2.32 N-*Dealkylation reactions*

Clearly, if a plant is able to metabolise a herbicide more quickly than the herbicide can accumulate at the site of action within the plant, then that plant will be tolerant of that herbicide. This is detoxification.

Some herbicides are applied in a form that is inactive as a herbicide and it is the plant's metabolic processes that converts the applied

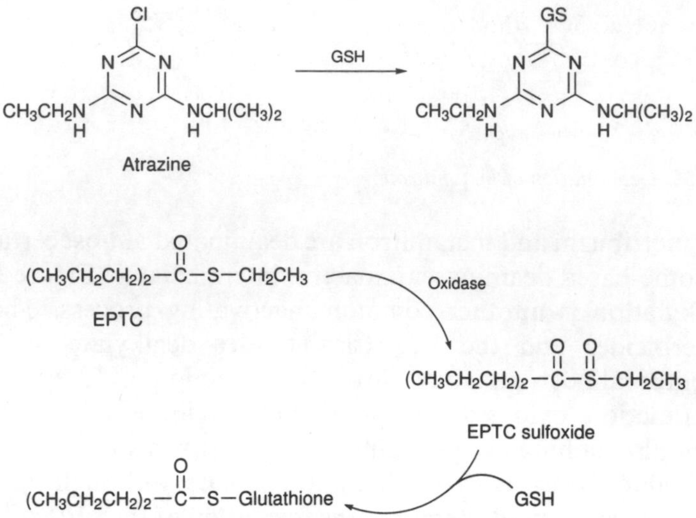

Figure 2.33 *Conjugation with glutathione*

compound into a herbicidal compound. Many acids are applied as esters, amides or salts and these have to be hydrolysed to the free acid before they become effective. Examples of this are found with the aryloxyalk-anoic acids, the aryloxyphenoxypropionates and the *N*-arylalanine esters (Figure 2.34). If these compounds are hydroxylated or detoxified in other ways before the free acid is released the compound will be ineffective as a herbicide on that weed or selective as a herbicide in that crop.

The oxidation of the thiocarbamates to form sulfoxides is another activating process as it has been shown that the sulfoxides are more

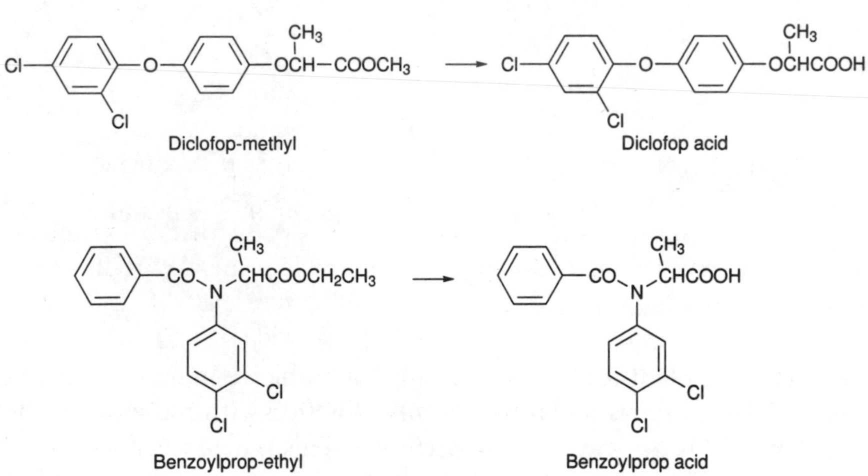

Figure 2.34 *De-esterification to release active free acids*

effective herbicides than the thio-compounds (Figure 2.33) but the sulfoxide also conjugates with glutathione, rendering it herbicidally inactive. The classical activation process is the β-oxidation of the herbicidally inactive MCPB into the herbicidally active MCPA (Figure 2.35), a process that takes place slowly in legumes and that, thereby, allows the use of compounds such as MCPB for the control of broad-leaved weeds in cereals undersown with clover.

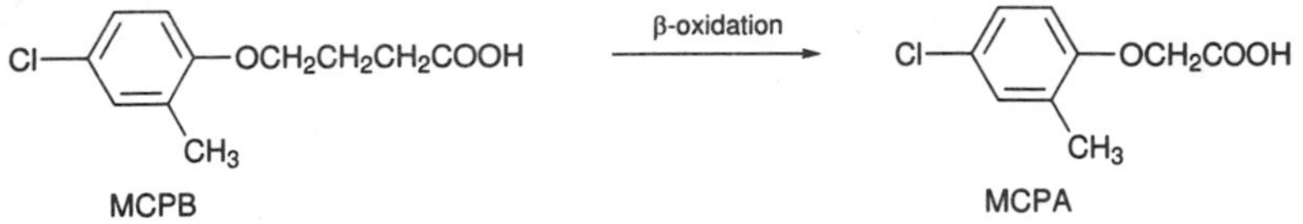

Figure 2.35 *β-Oxidation of MCPB to MCPA*

5 HERBICIDE RESISTANCE

Resistance to herbicides has developed more slowly than with insecticides or fungicides but it is now present and represents an increasing threat to herbicide efficacy. The first compounds to which significant resistance was noted were the 1,3,5-triazines. In this case there was a change in the triazine binding site in the chloroplast of resistant weeds such that the inhibitor (the herbicide) would not bind and the weed was unaffected. Fortunately, these weeds were often less fit than their susceptible counterparts and with the use of other herbicides with different modes of action they could be controlled. Other examples exist of weeds with changed target sites where resistance is becoming a serious problem primarily because the weeds are as competitive as the susceptible strains. The most serious at present are those weeds that are resistant to acetolactate synthase inhibiting herbicides. Many compounds in very wide use are active through this mode of action and it is usual for weeds resistant to one member to be resistant to all classes with the same mode of action.

A more serious problem is presented by weeds that are resistant because they are able to metabolise the herbicide, thereby reducing its concentration within the weed to an ineffective level. This type of resistance will often lead to resistance to a wide range of different herbicides with different modes of action.

In an attempt to help farmers prevent the onset of resistance in the weeds on their land the Herbicide Resistance Action Committee (HRAC) has produced a table of herbicide classes by mode of action (anon. 1996). This table is designed to tell the grower which compounds are related and how to select the best compound or compounds to prevent the onset of resistance or to reduce its impact once it is present (Table 2.2). If herbicides in one class are ineffective it is recommended that choices be made of herbicides from different classes or that

Table 2.2 *Summarised table of herbicide modes of action*

Group	Mode of action	Chemical family	WSSA[a] Group
A	Inhibition of acetyl CoA carboxylase (ACCase)	Aryloxyphenoxy-propionates Cyclohexanediones oximes	1
B	Inhibition of acetolactate synthase (ALS) [Acetohydroxy acid synthase (AHAS)]	Sulfonylureas Imidazolinones Triazolopyrimidines Pyrimidinylthiobenzoates	2
C_1	Inhibition of photosynthesis at photosystem II	1,3,5-Triazines Triazinones Uracils Pyridazinone Phenyl carbamates	5
C_2	Inhibition of photosynthesis at photosystem II	Phenylureas Amide	7
C_3	Inhibition of photosynthesis at photosystem II	Nitriles Benzothiadiazole Phenyl pyridazine	6
D	Photosystem-I electron diversion	Bipyridyliums	22
E	Inhibition of protoporphyrinogen oxidase (PPO)	Diphenyl ethers *N*-Phenylphthalimides Thiadiazoles Oxadiazoles Triazolinones	14
F_1	Bleaching: inhibition of carotenoid biosynthesis at the phytoene desaturase step (PDS)	Pyridazinones Nicotinanilides Others	12
F_2	Bleaching: inhibition of 4-hydroxyphenylpyruvate dioxygenase (4-HPPD)	Triketones Isoxazole Pyrazole	28

(Continued)

Table 2.2 *Continued*

Group	Mode of action	Chemical family	WSSA[a] Group
F_3	Bleaching: inhibition of carotenoid biosynthesis (unknown target)	Triazoles Isoxazolidinone Phenylurea	11 13
G	Inhibition of EPSP synthase	Glycines	9
H	Inhibition of glutamine synthetase	Phosphinic acids	10
I	Inhibition of dihydropterate synthase (DHP)	Carbamates	18
K_1	Microtubule assembly inhibition Phosphoroamidates	Dinitroanilines Pyridazines Benzoic acid	3
K_2	Inhibition of mitosis	Carbamates Benzylethers	23 27
K_3	Inhibition of cell division	Chloroacetanilides Carbamates Acetamides Benzamides Oxyacetamides	15
L	Inhibition of cell wall (cellulose) biosynthesis	Nitriles Benzamides	20 21
M	Uncoupling (membrane disruption)	Dinitrophenols	24
N	Inhibition of lipid biosynthesis – not ACCase inhibition	Thiocarbamates Phosphorodithioates Benzofurans Chloro-carbonic acids	8
O	Synthetic auxins	Phenoxyalkanoic acids Benzoic acids Pyridinecarboxylic acids Quinolinecarboxylic acids	4
P	Inhibition of indole-acetic acid action	Phthalamates	19
Z	Unknown	Arylaminopropionic acids Organoarsenicals Others	25 17 27 8

From Herbicide Resistance Action Committee Classification of Herbicides by Mode of Action.
[a]WSSA – Weed Science Society of America.

combinations of herbicides be used. More information can be obtained from Dr Robert Schmidt, Bayer, Monheim, Germany.

6 QUESTIONS

2.1 Why is it important to control weeds in crops?

2.2 What events drove the farmer towards the use of chemical crop protection?

2.3 List four biochemical targets for herbicides that might be considered to be good targets and three that would be poor targets. Why have you chosen these?

2.4 How would you show that a herbicide that inhibited photosynthesis required light to show its effect and that plant death was not due to starvation?

2.5 Why is ribulose bisphosphate carboxylase (RUBISCO) a poor target for a herbicide?

2.6 Why are some herbicides selective to some crops?

2.7 Is weed resistance a significant problem in many crops? How can it be combatted?

7 REFERENCES

1. J.A.R. Lockhart, A. Samuel and M.P. Greaves, 'The Evolution of Weed Control in British Agriculture', in 'Weed Control Handbook: Principles', 8th Edn., eds. R.J. Hance and K. Holly, British Crop Protection Council, Farnham, UK, 1990, pp. 43–74.

2. I. Ridge (ed.), 'Plant Physiology: Biology: Form and Function', Hodder and Stoughton for the Open University, Milton Keynes, UK, 1991.

3. Holy Bible, King James Version, Isaiah, Ch. 40, verse 6, 1611.

4. B.M. Luscombe and K.E. Pallett, 'Isoxaflutole for Weed Control in Maize', *Pesticide Outlook*, 1996, **7**(6), 29–32.

5. F.E. Dayan and S.O. Duke, 'Porphyrin-Generating Herbicides', *Pesticide Outlook*, 1996, **7**(5), 22–27.

6. K. Grossmann and F. Scheltrup, 'On the Mode of Action of the New, Selective Herbicide, Quinmerac', Brighton Crop Protection Conference – Weeds, 1995, British Crop Protection Council, Farnham, UK, pp. 393–398.

7. F. Scheltrup and K. Grossmann, 'Abscisic Acid is a Causative Factor in the Mode of Action of the Auxinic Herbicide, Quinmerac, in Cleaver (*Galium aparine* L.)', *J. Plant Physiol.*, 1996, **147**, 118–126, .

8. M.P. Greaves, 'Microbial Herbicides – Factors in Development', in 'Crop Protection Agents from Nature: natural products and analogues', ed. L.G. Copping, The Royal Society of Chemistry, Cambridge, UK, 1996, pp. 444–467.

9. C.J. Cilliers, 'Weed Control with Insects – a Matter of Fitness', in

'Comparing Glasshouse and Field Pesticide Performance II, BCPC Monograph 59', eds. H.G. Hewitt, J.C. Caseley, L.G. Copping, B.T. Grayson and D. Tyson, British Crop Protection Council, Farnham, UK, 1994.
10. C.D.S. Tomlin (ed.), 'The Pesticide Manual', 11th Edn., British Crop Protection Council, Farnham, UK, 1997.

CHAPTER 3

Insecticides

1 INTRODUCTION

Man has battled against insects ever since he first grew arable crops to feed and shelter his family and to nourish his livestock. The devastation that can be caused by insects is recounted in the Bible with two of the plagues of Egypt: "*Else, if thou wilt not let my people go, behold I will send swarms of flies upon thee, and upon thy servants, and upon thy people, and into thy houses: and the houses of the Egyptians shall be full of swarms of flies . . . . the land was corrupted by reason of the swarm of flies*" and "*Else, if thou refuse to let my people go, behold, to morrow will I bring the locusts into thy coast: And they shall cover the face of the earth, that one cannot be able to see the earth: and they shall eat the residue of that which is escaped, which remaineth unto you . . . . and shall eat every tree which groweth for you out of the field. . . . . . . For they covered the face of the whole earth, so that the land was darkened; and they did eat every herb of the land, and all the fruit of the trees: and there remained not any green thing in the trees, or in the herbs of the field, through all the land of Egypt.*"[1]

Not surprisingly, Pharaoh let the people of Israel go!!

Today insecticide usage is second only to herbicides in terms of global sales, but in the developing world, insecticides are the largest investment in crop protection chemicals. It is estimated that insect pests cause losses of between 13 and 16% of yield with a value of about $90 billion.[2] In 1961, it was estimated that there were 9 000 species of insect and mite that infest crops and most of these are naturally occurring insects that have moved from native vegetation onto introduced crops.[3] A very good example of this is the Colorado potato beetle that was naturally occurring in the western United States and became a significant problem as potatoes were introduced. It is now present across the entire country, causing significant damage to the crop. It is reassuring to know, however, that despite these large numbers of potentially damaging insect and mite species, relatively few cause significant damage to crops.

In traditional agricultural systems with sufficient labour to work the land it was relatively easy to control weeds through their manual removal. Insects, however, are clearly damaging to the crop and are much more difficult to control by hand removal. There are examples of teams of workers searching through cotton crops to remove the egg masses of lepidopterous pests but these are the exception rather than the rule. For this reason, chemicals have been applied to crops to control insect pests for over 100 years. The early compounds were used as winter washes as the ingredients were general toxicants and would kill the crop as well as the insect if applied when the crop was growing actively. The winter wash was designed to be applied to perennial crops such as orchards over the winter and preferably just before bud burst to kill the eggs of phytophagous insects before they had the opportunity to infest the crop, damage the fruit and reproduce. Oils, detergents, lime sulfur and dinitrophenols were typical examples of the type of compound that was applied.

It was the discovery of DDT during the Second World War that revolutionised insect control in agriculture. The compound was broad-spectrum, insect specific, non-phytotoxic and persistent – the ideal compound to use on all sorts of crops to control these damaging species. Today, DDT has a poor reputation, and it is clear that if it were discovered today it would not be developed, but the compound was invaluable as a means of controlling pests on crops immediately after the war and, perhaps more importantly, it was used to delouse soldiers, peasants and prisoners across Europe thereby saving millions of lives. Its value as a means of controlling diseases that were vectored by insects is also often underestimated. Malaria is still the biggest killer of people in the world. The use of DDT to control the mosquito vector of the parasite led to a dramatic reduction in the incidence of the disease throughout the world and it is claimed that, had the WHO spray programme been continued, malaria would have been eradicated.

Control of insects and mites will be discussed together in this chapter.

2 BIOCHEMICAL MODES OF ACTION

The control of insects has, traditionally, been associated with interference with nerve function. This makes many insecticides relatively toxic to non-target organisms, and particularly beneficial insects and mammals, including man. Selectivity will not be reviewed in any detail as this is associated with a number of factors such as penetration to the target site and different rates of metabolism in different organisms. The review of the biochemical mode of action of insecticides in this chapter will describe the nervous system of insects in brief and then discuss how different classes of insecticide exert their insecticidal effects. The later sections of the chapter will review other target sites and non-cidal methods of insect control.

2.1 Insect Nervous System Disruption

The target sites within the nervous system of insects known at present are very restricted. They consist of the sodium channel, the components of the nicotinic cholinergic synapse and the γ-aminobutyric acid (GABA) and octopamine receptors. Benson[4] (1991) lists potential target sites within the insect neuronal and muscular system. These are shown in Table 3.1.

Nerve function is a transfer of electrical pulses through nerve cells and across the gaps between nerve cells, the synapse, so that a message is transmitted from the brain to a muscle or other responsive tissue or from sensory tissue to the brain. It is possible to interfere with nerve function in a number of ways. Nerve impulses pass down the axons of nerve cells (long processes of each nerve cell) as a result of changes in the permeability of the axon membrane to sodium and potassium ions. When at rest, the electrical potential within the membrane is negative in comparison to the outside. The concentration of sodium inside the cell is low and the concentration of potassium is high. Potassium and sodium ions enter the cell using two mechanisms: sodium and potassium channels (or gates) allow a rapid passive movement when opened, whilst a slower, active movement occurs through ion pumps. A nerve impulse passing down an axon is a wave of changing polarity that is caused by the sodium gate opening so that sodium passes in and then the potassium gate opening so that potassium can move out, thereby restoring the electrical polarity. The resting condition is restored by the operation of the ion pump taking up potassium at the expense of ejected sodium.

At the end of an axon, where it meets another nerve cell or an effector cell (a cell such as a muscle or a gland cell), there is a gap or junction that is usually about 10 to 20 nm wide and this is known as a synapse. The passage of the nerve impulse across this synapse is chemical rather than electrical. When the nerve impulse reaches a synapse it causes the release of a chemical transmitter that is usually acetylcholine. Other transmitters have been identified and these include L-glutamate and γ-aminobutyric acid (GABA). The released acetylcholine interacts with a receptor on the

Table 3.1 *Potential neuronal and muscular insecticide target sites*

Target	In vivo *activator*	Commercial insecticide class
Neurotransmitter receptor ligand recognition sites		
Cholinergic		
Nicotinic	Acetylcholine	Nicotine
		Cartap
		Nitromethylenes
Muscarinic	Acetylcholine	None
Glutaminergic	Glutamate	None
Octopaminergic	Octopamine	Amitraz
GABAergic	γ-Aminobutyric acid	None
Ion channels		
Na$^+$ channel	Depolarisation	Pyrethroids
		DDT
Cl$^-$ channel	γ-Aminobutyric acid	Cyclodienes
GABA regulated		Avermectins
		Fipronil
Secondary messenger systems		
Cyclic AMP	Octopamine, 5HT	None
Transmitter re-uptake and breakdown systems		
Cholinesterase	Acetylcholine	Organophosphates
		Carbamates
Mitochondrial respiration		
Oxidative phosphorylation	–	Rotenone
		Dinitrophenols
		Diafenthiuron
Muscle		
Contraction	Depolarisation	Ryania extract

adjoining cell and the binding of the acetylcholine with this receptor causes this post-synaptic cell to pass on the impulse (if it is another axon) or to do work (if it is a muscle or gland cell). If this chemical signal were not controlled then the message would continue to be transmitted without the electrical stimulation from the axon. The control is achieved by the presence of acetylcholinesterase, an enzyme that hydrolyses the acetylcholine and thereby prevents the continual surge of signals and frees the receptor to receive another signal. This is represented diagrammatically in Figure 3.1.

The site of action of all organophosphorus and carbamate insecticides

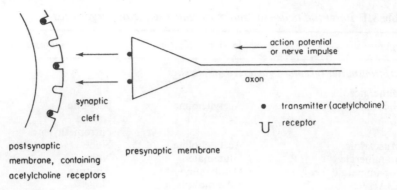

Figure 3.1 *Diagrammatic representation of the functioning of a nerve synapse*

is this enzyme, acetylcholinesterase, that hydrolyses the synaptic transmitter, acetylcholine.

$$(CH_3)_3N^+CH_2CH_2OC(O)CH_3 + H_2O \rightleftharpoons (CH_3)_3N^+CH_2CH_2OH + CH_3COOH$$

A wide range of insecticides (and acaricides and nematicides) show this mode of action. In the 11th Edn of 'Pesticide Manual' there are 71 organophosphorus and 18 carbamate insecticides listed.[5]

2.1.1 Organophosphorus Insecticides (OPs). It is often said that OP insecticides are derived from nerve gases and we once worked for an agrochemical company that had been told by its research director not to make or test compounds containing phosphorus because such research would reflect badly on the company's other activities. Whilst it is true that many of today's insecticides are effective through an interference with nerve function it is over simplistic to group insecticides and agents of war together. However, insects and mites are animals and animals move, feed, breed and interact as a result of the detection of stimuli and the ability to respond to these stimuli. This is accomplished through the transfer of messages through the nervous system with the resulting associated reaction. If these messages are interrupted or otherwise disturbed, the necessary reaction will be impaired or prevented and the insect will die. Consequently, many insecticides interfere with nerve function. Clearly, there is a possibility that this mode of action will have an effect on other non-target animals, including man, and so it is important to develop only those compounds that are selectively active against insects and mites.

The OPs are very significant insecticides in global crop protection.

There are several different substitution patterns within the group and these are shown in Figure 3.2.

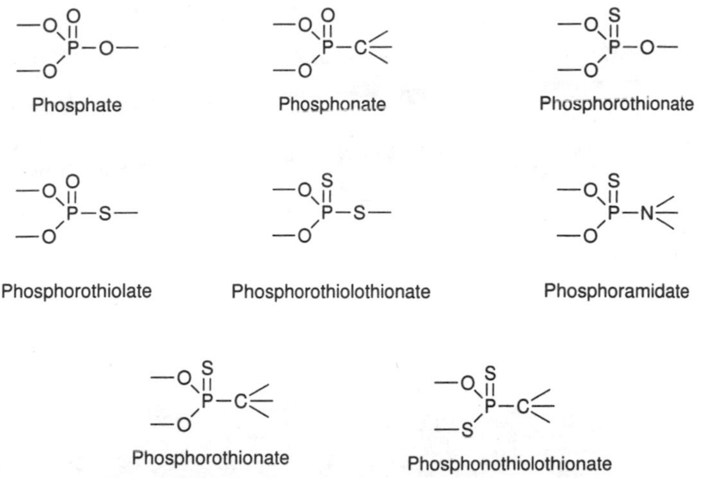

Figure 3.2 *Organophosphorus nomenclature*
(From J.R. Corbett, K. Wright and R.C. Baillie, The Biochemical Mode of Action of Pesticides, Academic Press, London, 1984)

It is important to know that the inhibition of acetylcholinesterase by OPs is through an attack on the relatively positive phosphorus atom by the hydroxyl group of a serine residue at the enzyme's site of action. Electron withdrawing substitutions within the OP tend to make the phosphorus more positive and, therefore, more reactive. Unfortunately, this type of substitution also makes the compound less stable hydrolytically. The discovery and development of OP insecticides has always been a balance between activity against the enzyme of the insect, selectivity in comparison with mammalian systems and stability within the insect. The binding of OPs to acetylcholinesterase is often irreversible. Typical OP insecticides are shown in Figure 3.3.

The importance of this group of compounds is reflected by the fact that chlorpyrifos has been the world's largest selling insecticide (in tonnage terms) for the last ten years.

2.1.2 Carbamate Insecticides. Carbamate insecticides interact with acetylcholinesterase in exactly the same way as OPs, with the hydroxyl group in the serine at the enzyme's active site attacking the carbamate residue in the insecticide. However, the binding to the active site is reversible. Typical carbamate insecticides are shown in Figure 3.4.

Both OPs and carbamates inhibit the active site of the target enzyme. This results in the uncontrollable firing of neurons, leading to loss of

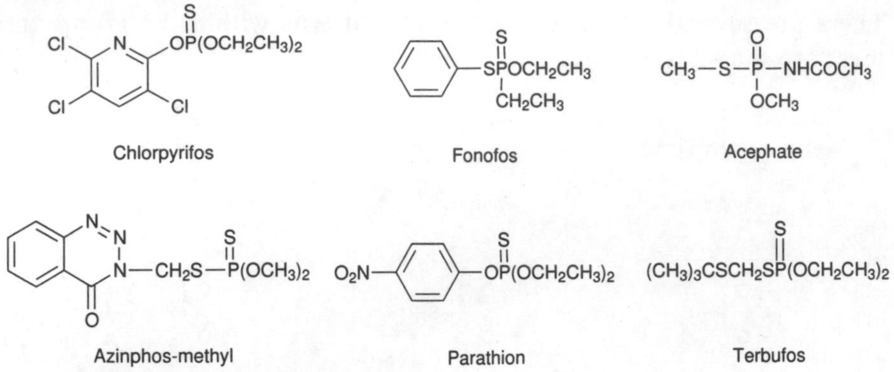

Figure 3.3 *Typical OP insecticides*

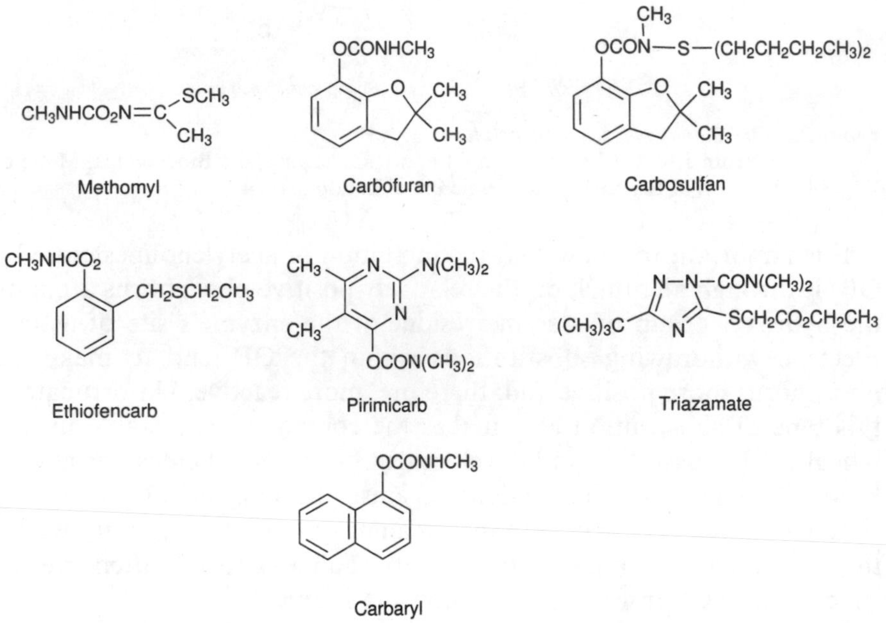

Figure 3.4 *Carbamate insecticides*

coordination and a massive release of hormones, resulting in water loss and death.

2.1.3 Compounds that Interact with Neurotransmitter Ligand Recognition Sites. The OPs and carbamates interfere with the enzyme that hydrolyses the chemical messenger, acetylcholine, thereby preventing a new electrical pulse being detected by a new surge of acetylcholine. Clearly, if the ligand to which the chemical messenger binds is filled with an inhibitor that mimics the normal acetylcholine transmitter, then the binding of these inhibitors to the receptor will lead to uncontrollable firing of neurons. Several compounds exert their insecticidal effects in this fashion. There are two cholinergic acetylcholine receptors termed nicotinic and muscarinic. The names are derived from inhibitors that were originally found to block the receptor nicotinic from nicotine, an alkaloid from tobacco, *Nicotiana tabacum*, and muscarinic from muscarine, an alkaloid from the fly agaric, *Amanita muscaria*.

Nicotine (Figure 3.5) has been known for over 300 years to be insecticidal[6] and was used as a foliar spray and as a vapour to control insects, particularly in glasshouses and other covered crops. The toxin nereistoxin (Figure 3.5), derived from the marine worm (*Lumbriconereis heteropoda*), has been modified to produce an analogue, cartap (Figure 3.5), that is converted into nereistoxin in the insect, leading to its death. An analogue of cartap, bensultap, was developed later with reduced mammalian toxicity. All of these compounds exert their toxicity by binding to the acetylcholine receptor and mimicking the effect of acetylcholine.

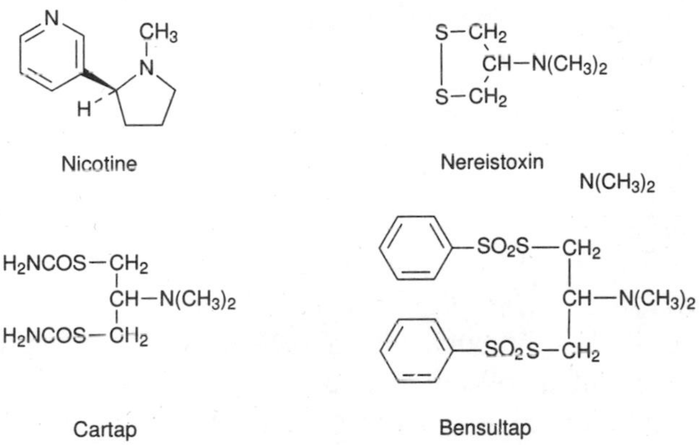

Figure 3.5 *Nicotine, nereistoxin, bensultap and cartap*

A recently introduced insecticide from the nitromethylene group, imidacloprid (Figure 3.6), has been shown to work in exactly the same way as nicotine and cartap but is of more interest commercially because

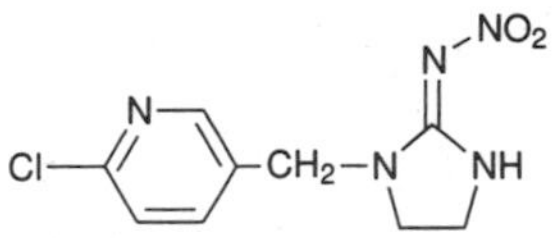

Figure 3.6 *Imidacloprid*

it has much reduced mammalian toxicity and because it has systemic properties. This means that the compound can be applied to the soil around infested plants and it will be taken up by the plants and control the insects that are infesting it. This is a valuable property for the control of insects such as aphids that feed on the phloem. It should be emphasised, however, that imidacloprid controls both sucking and chewing insects and is effective through the stomach and by contact action.[7]

Although their are some natural inhibitors of the muscarinic receptor, no insecticidal product has been commercialised yet with this mode of action.

In addition to acetylcholine there are three other neurotransmitters in animal nervous systems. These are glutamate, octopamine and GABA. Surprisingly, only inhibition of the octopaminergic receptors has led to the introduction of a product. This is amitraz (Figure 3.7), a compound that is unusual in that it is a very effective acaricide, with additional effects on cattle ticks, but it also shows activity against lepidopteran insect eggs and against selected homopteran insect species, particularly pear sucker, *Psylla pyricola*, and cotton whitefly, *Bemisia tabaci*. Related formamidine insecticides have now been withdrawn from the market.

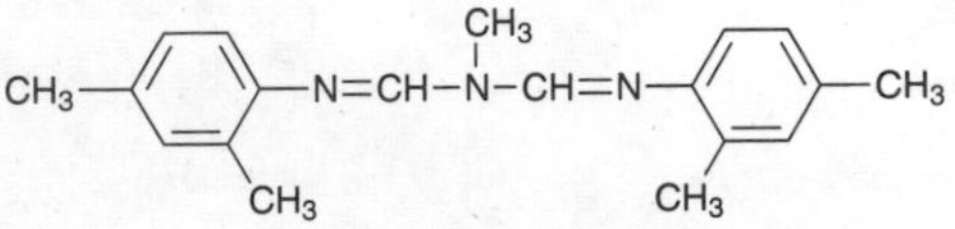

Figure 3.7 *Amitraz*

2.1.4 Insecticides that Interfere with Ion Channels. DDT (Figure 3.8) was introduced as an insecticide in the 1940s and following its introduction a large number of chlorinated hydrocarbon insecticides were developed and marketed. These compounds were responsible for a revolution in insect control methods but have lost their importance because of their unacceptably long persistence in the environment, their fat solubility, which meant that they accumulated in fatty tissue of non-target organisms, and because of the onset of insect resistance. DDT binds to the sodium channel of the insect's nervous system, causing leakage and thereby preventing the electrical pulse from moving through the axon.

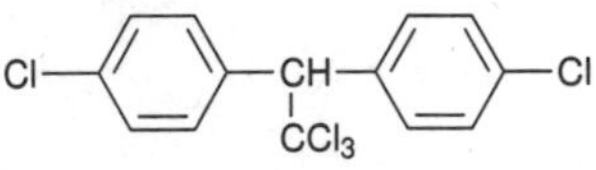

Figure 3.8 *DDT*

One of the most successful insecticide groups ever introduced is the synthetic pyrethroids (Figure 3.9); these compounds have been shown to

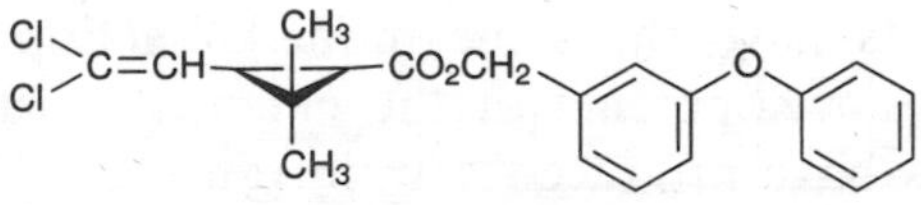

Permethrin

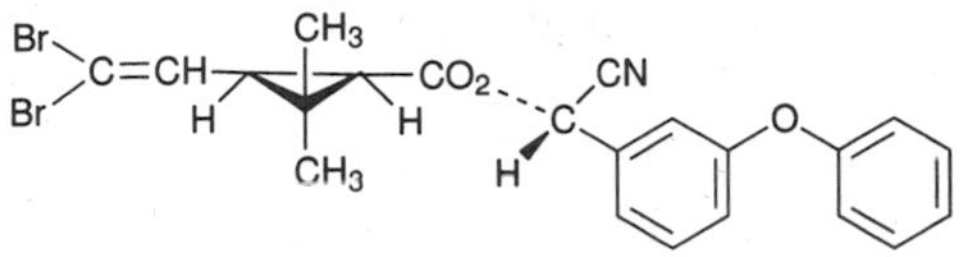

Deltamethrin

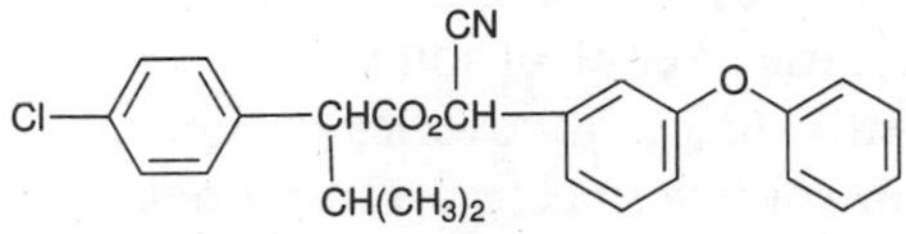

Fenvalerate

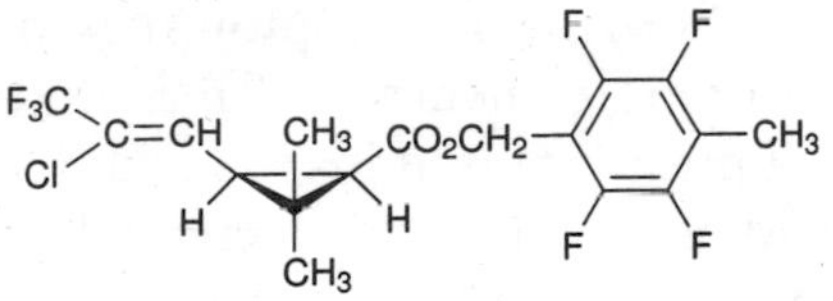

Tefluthrin

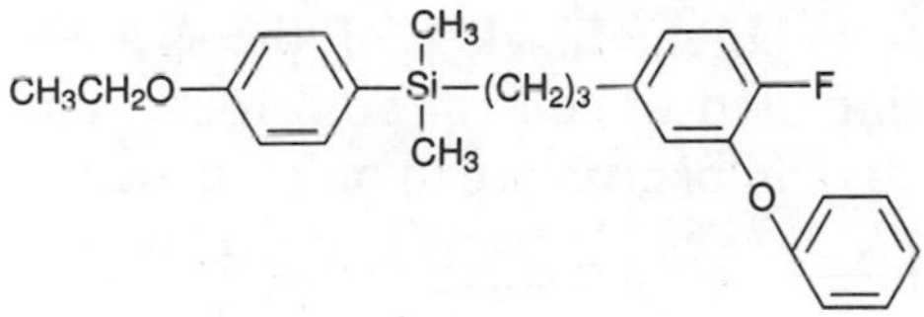

Silafluofen

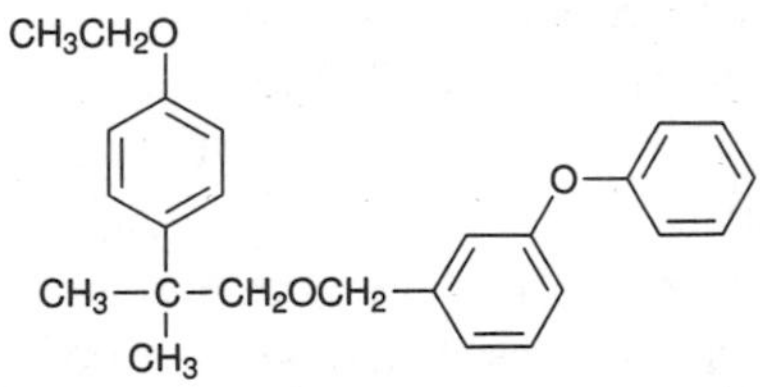

Etofenprox

Figure 3.9 *Some synthetic pyrethroids*

bind to the sodium channels in insects, prolonging their opening and thereby causing knockdown and death. (For an outstanding account of the history and development of synthetic pyrethroids see Elliott.[8]) The synthetic pyrethroids have the same mode of action as the natural pyrethrins. Recent developments in the chemical structures of these synthetic analogues have produced a wide range of related but novel chemical classes. All are characterised by good, broad-spectrum insecticidal activity with some showing effects against mites. Use rates are usually low and this increases the safety of the compounds to non-target organisms. In addition, the compounds are bound in the soil, rendering them unavailable to soil insects, are generally non-volatile, reducing their drift, and possess insect repellent activity, thereby reducing their impact upon beneficial insects. More recent substitutions have led to the synthesis of non-ester pyrethroids (*e.g.* etofenprox), compounds containing silicon (silafluofen) and compounds with good volatility characteristics, allowing their use to control soil inhabiting insects (tefluthrin).

The target for two major discoveries within microbial products for insect and mite control (avermectins and milbemycins – Figure 3.10) is the GABA receptor in the peripheral nervous system. Both classes of compound stimulate the release of GABA from nerve endings and enhance the binding of GABA to receptor sites on the post-synaptic membrane of inhibitory motor neurons of nematodes and on the post-junction membrane of muscle cells of insects and other arthropods. This enhanced GABA binding results in an increased flow of chloride ions into the cell, with consequent hyperpolarisation and elimination of signal transduction resulting in an inhibition of neurotransmission.

More recently, the new insecticide, fipronil (Figure 3.11), has been shown to act as a potent blocker of the GABA-regulated chloride channel. It is being used to control both foliar and soil insects[9] whilst the avermectins and milbemycins can only be used against foliar pests.

2.2 Inhibition of Oxidative Phosphorylation

Many of the earliest insecticides exerted their effects through an inhibition of oxidative phosphorylation, they were uncouplers. This meant that they uncoupled the electron transport chain from the production of ATP (the formation of chemical energy). 2,4-Dinitrophenols, such as dinitro-*o*-cresol (Figure 3.12), were effective as winter washes and later were developed as herbicides because compounds of this type were general toxicants, causing death to most living organisms that they encountered. The natural insecticide rotenone, found in the plant genera *Derris*, *Lonchocarpus* and *Tephrosia*, interferes with respiration

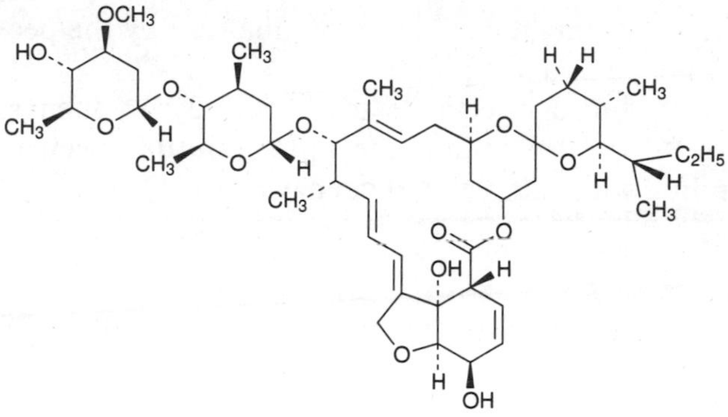

Avermectin A₁ₐ

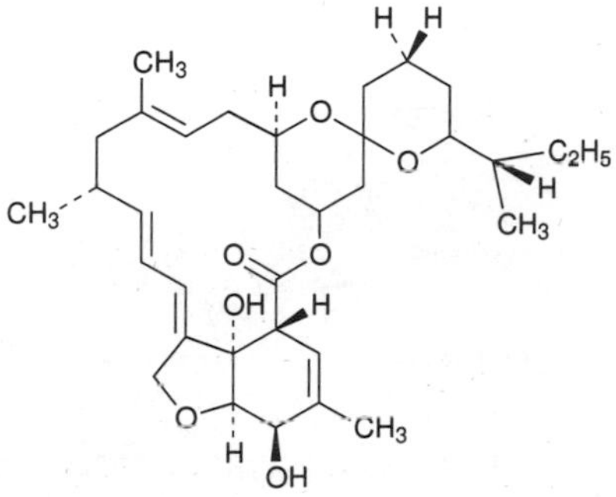

Milbemycin D

Figure 3.10 *Avermectin A₁ₐ and milbemycin D*

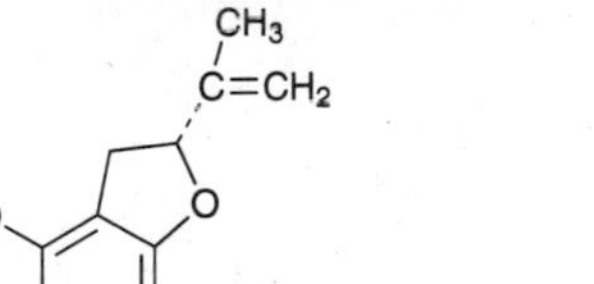

Figure 3.11 *Fipronil*

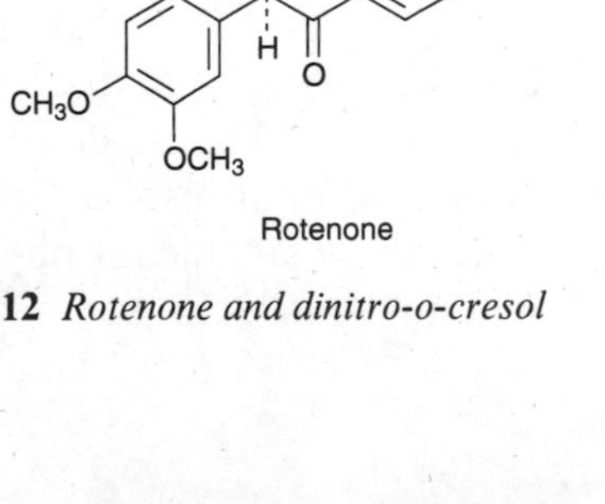

Rotenone

Dinitro-*o*-cresol

Figure 3.12 *Rotenone and dinitro-o-cresol*

at site I, a fact that might explain its original use by the people of Asia and South America as a fish poison.

Recently introduced insecticide/acaricides, pyrimidifen and fenaza-quin (Figure 3.13), also inhibit the mitochondrial electron transport chain by binding with complex I at coenzyme site Q.

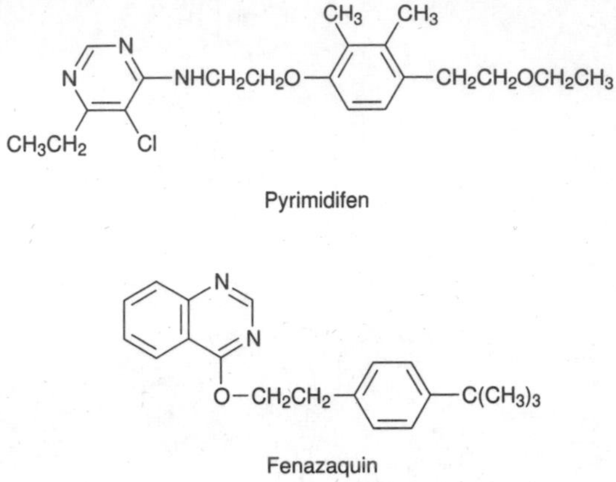

Pyrimidifen

Fenazaquin

Figure 3.13 *Pyrimidifen and fenazaquin*

A recently introduced insecticide/acaricide, chlorfenapyr (Figure 3.14), has been shown to disrupt the electrochemical gradient in mitochondria and, thereby, uncouple oxidative phosphorylation. There is also evidence that the compound is a pro-insecticide, being converted into the active form by mixed function oxidases within the insect or mite.

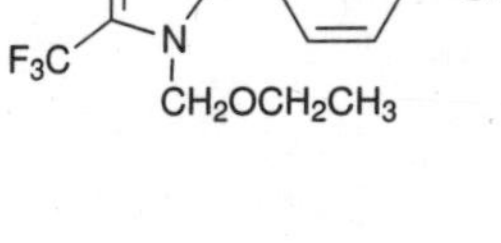

Figure 3.14 *Chlorfenapyr*

More recently, a new class of insecticides derived from naphthoqui-nones found in the South American alpine plant *Calceolaria andina* has entered development. Compounds are being developed jointly through the British Technology Group (BTG) (following the discovery at Rothamsted Experimental Station) and by Bayer. The mode of action of these compounds is believed to be an inhibition of respiration at site III. Consequently, some of the compounds show significant phytotoxi-city to crops. Typical structures are shown in Figure 3.15.

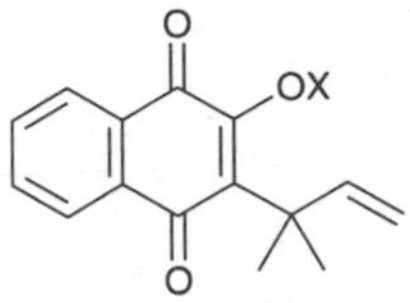

BTG 504 X = Acetate
BTG 505 X = H

Figure 3.15 *Naphthoquinones with insecticidal activity*

It is interesting that a new group of fungicides based on the natural products from the fungus *Strobilurus tenacellus* also inhibit mitochondrial respiration at the site of complex III (bc_1-complex) of the respiratory chain (see Chapter 4). Recently synthesised compounds from within this class are showing interesting insecticidal effects.

2.3 Insect Growth and Regulation

Insects pass through a number of developmental stages from egg to adult. As a general rule the larval stages are targeted at rapid growth through feeding whilst the adult stages are involved in reproduction. For this reason it is usually the larval stages that are the crop destructive segments of the insect's life cycle and it is these that are targeted by insecticides. This is clearly not the case with vectors of human or animal parasites such as malaria, sleeping sickness and Dengue fever where it is the adult that transmits the parasites and these that are attacked in protection strategies. Other insects, such as aphids, give birth to live young and, although the young shed their cutinous skin as they expand and grow, they do not have the classical life cycle stages that are typical of many phytophagous pests, including lepidopteran, coleopteran, hemipteran and acarinal species.

There are a number of complex developmental processes that can be interrupted within the growth cycle of most insect pests and a number of compounds have been developed to exploit these. The benzoylurea insecticides (Figure 3.16), sometimes known as insect growth regulators, are compounds that interfere with chitin biosynthesis in the insect. Hence, as a larva prepares to moult and replace its shed outer skin with a newly synthesised replacement, there is insufficient chitin available to complete the construction of the outer layers and the larva dies either during or immediately after moulting. The name 'insect growth regulator' is probably derived from the often distorted shapes of dead and dying insects treated with these compounds.

Some of these compounds are active against a wide range of insect and mite species but the most sensitive are those that grow rapidly and moult

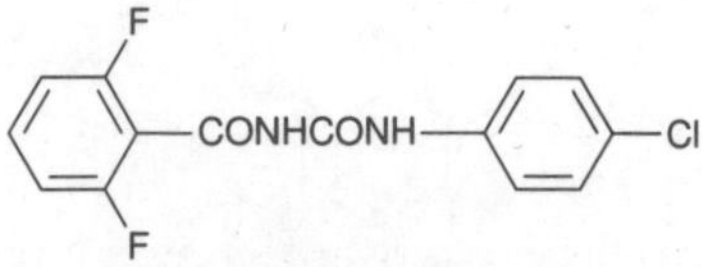

Diflubenzuron

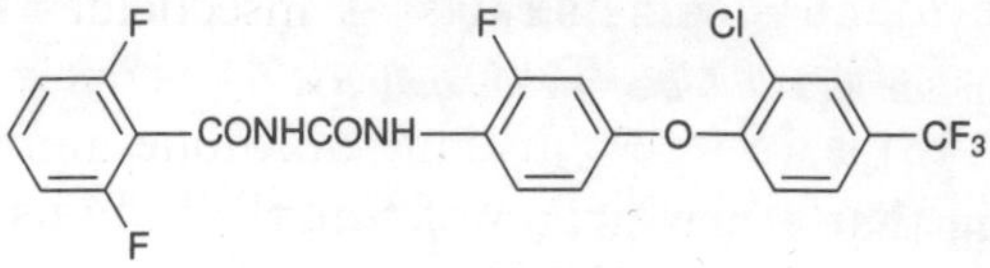

Flufenoxuron

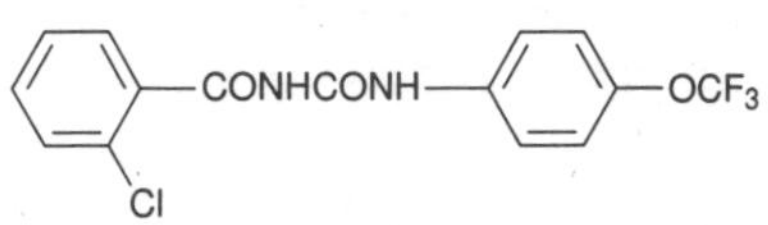

Triflumuron

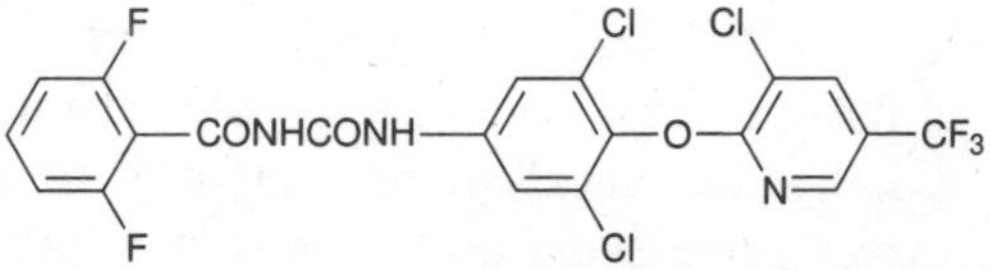

Chlorfluazuron

Figure 3.16 *Benzoylurea insecticides*

frequently, such as Lepidoptera. Because the insects have to take the compound up and inhibition is not effected until moulting, the compounds are relatively slow acting in comparison with neural toxins. However, because they have poor contact activity and have to be consumed to be effective, they are relatively safe to non-target and beneficial insect species. Consequently, these compounds are widely recommended as components of Integrated Crop Management systems. The insecticide/acaricide buprofezin (Figure 3.17), has the same mode of action but is unsual in its spectrum of activity being particularly effective against whitefly, *Bemisia* sp. and *Trialeurodes* sp. In addition, the

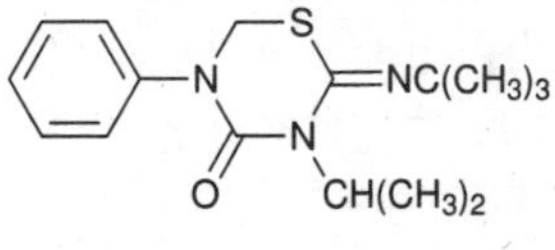

Buprofezin

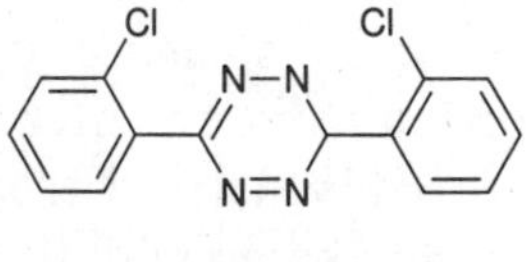

Clofentezine

Figure 3.17 *Buprofezin and clofentezine*

acaricide clofentezine (Figure 3.17) is thought to have the same mode of action as the benzoylureas as it shows cross-resistance to benzoylurea-resistant mites. This has yet to be demonstrated directly.

Insect growth hormones have been studied as possible targets for new insecticides but have generally been found to be ineffective. Some hormones have found small markets as insecticides but this has not proved to be a useful source of commercially viable products. Methoprene (Figure 3.18) is an insect juvenile hormone mimic and interrupts the normal development of adults when applied to larval stages. Such a compound would tend to prolong the crop destructive larval stages of many phytophagous insects and, hence, shows little promise as a crop insecticide. It is used to control fleas and several stored product pests as well as ants and mosquitoes.

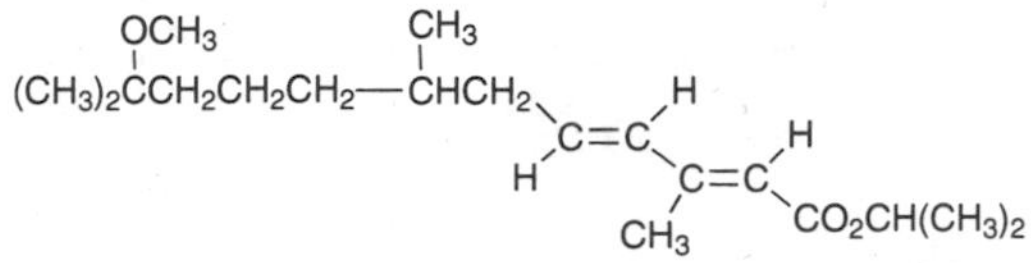

$$(CH_3)_2CCH_2CH_2CH_2-CHCH_2 \quad \text{(with } OCH_3 \text{ and } CH_3 \text{ substituents)}$$

$$C=C, \quad C=C, \quad CH_3 \quad CO_2CH(CH_3)_2$$

Figure 3.18 *Methoprene*

Many insects communicate with the release and detection of volatile compounds known under the general title of pheromones. These compounds are used to find a suitable food source, to alert other members of the species of potential danger and to find a mate. Mating pheromones were identified as very useful compounds to disrupt mating and thereby reduce insect populations through the failure of females to lay fertile eggs. It is usual for the fecund unfertilised female to release a sex pheromone that will attract males who fertilise the female. Subsequently, the female lays her eggs. These pheromones are volatile compounds that can be detected by the adults at low concentrations over relatively long distances. They are also species specific. Attempts to release pheromones such that mating is disrupted through an amplification of the pheromone concentration, leading to confusion of the adult males, has been partially successful in that the pink bollworm, *Pectinophora gossypiella*, pheromone (Figure 3.19) has been commercialised and is used successfully to disrupt the mating of this important cotton pest.

Sex pheromones have found a place in monitoring for the presence of insects as a guide to when spraying of conventional chemicals might be needed. However, the use of pheromones to attract insects and then trap them is not an effective control strategy because males only are attracted and the trapping has to be close to 100% effective to reduce mating in the field and this, to date, has been unachievable. The use of 'lure and kill'

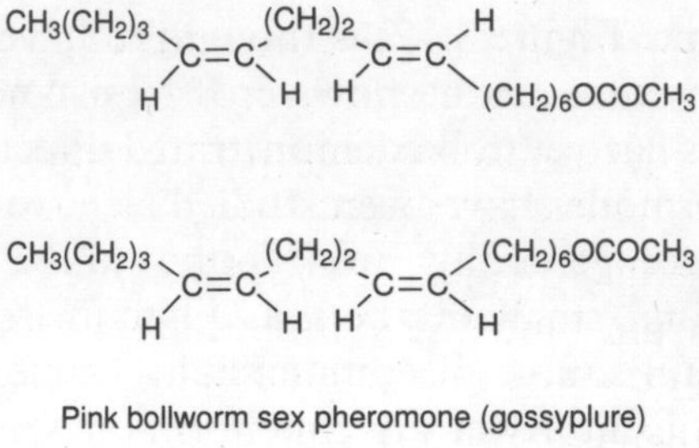

CH₃(CH₂)₃, (CH₂)₂, H
C=C, C=C,
H H H (CH₂)₆OCOCH₃

CH₃(CH₂)₃, (CH₂)₂, (CH₂)₆OCOCH₃
C=C, C=C,
H H H H

Pink bollworm sex pheromone (gossyplure)

Figure 3.19 *Pink bollworm sex pheromone*

pheromone traps is moderately effective in some situations. Here the males are attracted by a sex pheromone and are immediately treated with an insecticide or an entomopathogen (see later) that will kill them. In practice, this is an improved trapping device.

3 OTHER INSECT CONTROL STRATEGIES

Several compounds are used as insecticides where little is known about their biochemical mode of action or whether they are general toxicants. Some have been around for many years, some are of natural origin and others are living organisms that predate on insects.

3.1 Compounds with Uncertain Modes of Action

Tin-containing compounds have been used in agriculture for many years as both fungicides (Chapter 4) and as acaricides. Tricyclohexyltin hydroxide and several related compounds (Figure 3.20), are very active against adult spider mites and are believed to act through an inhibition of oxidative phosphorylation.

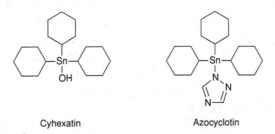

Cyhexatin   Azocyclotin

Figure 3.20 *Tricyclohexyltin hydroxide and azocyclotin*

4 NATURALLY OCCURRING COMPOUNDS

4.1 Micro-organisms as a Source

Several insecticides have been derived from naturally occurring compounds that show useful biological effects. Examples from plants already cited include pyrethrum (and the synthetic pyrethroids), nicotine and rotenone, whilst micro-organisms encompass avermectins, milbemycins and strobilurins. Other potential modes of action have been identified from naturally occurring compounds. For example, the cyclopolylactone tetranactin (Figure 3.21), shows good activity against spider mites and is effective as an ionophore interfering with membrane function, causing leakage of basic cations, such as potassium ions, through the lipid layer of the mitochondrial membrane. This type of compound requires the presence of water to be effective. The amino acid analogue alanosine (Figure 3.21) is effective against lepidopteran larvae through inhibition of insect ecdysis. Inhibition of the enzymes necessary for the metabolism of essential carbohydrates has also been demonstrated with natural products such as the aminosugar derivative allosamidin (Figure 3.21) which disrupts chitinase activity, and trehalostatin, which interferes with the activity of trehalase, an enzyme responsible for the hydrolysis of the storage carbohydrate, trehalose.

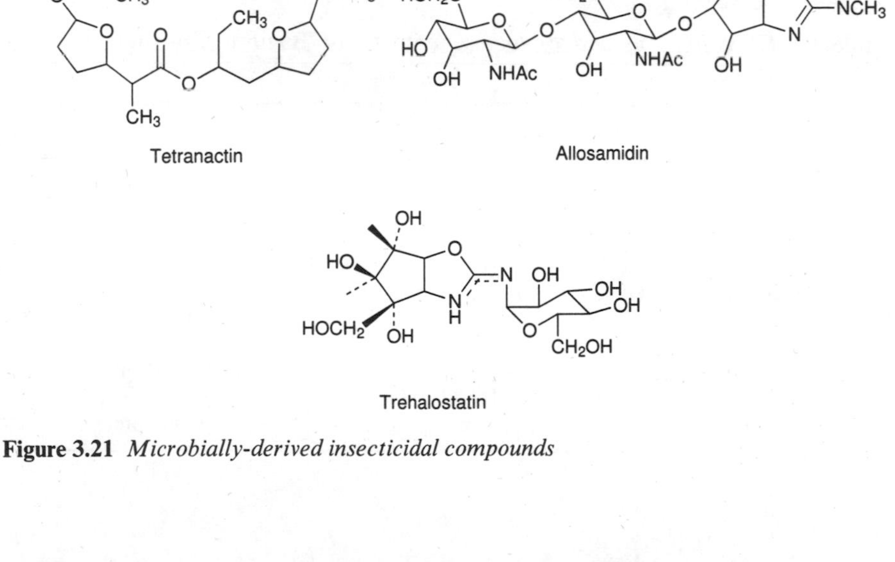

Tetranactin

Alanosine

Allosamidin

Trehalostatin

Figure 3.21 *Microbially-derived insecticidal compounds*

The most significant microbial insect active compounds are the delta-endotoxins from the motile, Gram-positive, soil-inhabiting bacterium *Bacillus thuringiensis* (*Bt*). During sporulation, *Bt* produces parasporal crystal inclusions that are insecticidal upon ingestion to susceptible species from the insect orders Lepidoptera, Diptera and Coleoptera. The inclusions vary in number, shape (the most common is bipyramidal) and composition and comprise one or more proteins – delta-endotoxins. The actual mode of action is not completely understood but involves the activation of the toxin by insect gut proteases, transport to the insect's peritrophic membrane, brush-border membrane binding and finally the formation of pores in the gut membrane that lead to exposure of the insect's haemolyph and mortal infection of the insect. The nature of pore formation is an area of research that remains largely obscure. It is possible that the pore is formed by the *Bt* toxins alone, a combination of the *Bt* toxin and the receptor or that the *Bt* toxin alters the receptor in such a way as to lead to the formation of a pore. It has even been speculated that the *Bt* toxin signals a secondary messenger system that is related to pore formation. Work is in progress in a number of laboratories to isolate and characterise the receptor molecule to allow these different possibilities to be clarified.

Different subspecies have been shown to be active against different orders of insect. Although the method of insect death is the same, the different endotoxins either have different binding regimes or are activated by the gut enzymes of different insect species and this leads to the differences in activity spectrum. There is evidence, however, that different insect orders act upon the delta-endotoxins in different ways. Treatment of a pro-toxin with lepidopteran gut juice leads to a toxin that is active against Lepidoptera whereas treatment with dipteran gut juice produces

Table 3.2 *Classification and biological activity of* Bacillus thuringiensis *subspecies*

Subspecies	Cry proteins	Biological spectrum
Bt kurstaki (*Btk*)	Cry1Aa, Cry1Ab Cry1Ac	Lepidoptera
Bt aizawai (*Bta*)	Cry1Aa, Cry1Ac, Cry1C, Cry1D, Cry1G	Lepidoptera
Bt tenebrionis (*Btt*)	Cry1Ac, Cry3A	Coleoptera
Bt israelensis (*Bti*)	Cry4A, Cry4B, Cry4C Cry4D	Diptera
Btk/Bta transconjugant	Cry1Aa, Cry1Ac, Cry1C, Cry1D	Lepidoptera
Btk/Btt transconjugant	Cry1Ac	Coleoptera

a toxin active against Diptera. A rough classification of the subspecies, the endotoxins (represented as Cry proteins) they produce and the insecticidal spectrum of each is shown in Table 3.2. For a sound review of *Bt* activity, structure and function see Adams *et al.*[10]

As these toxins are proteins it is possible to isolate the genes that code for them and use these genes to transform crop plants and thereby render them insecticidal. This is discussed briefly under Transgenic Crops (Section 6).

4.2 Higher Plants as a Source

In addition to the pyrethrums, there are a great number of higher plant derived compounds that show insecticidal activity. Before becoming very excited about these compounds it must be remembered that biologically active compounds in plants are usually defence mechanisms and, to the plant, any organism that feeds off it is a predator be that organism an insect or mite, a rabbit or a human. Hence, many natural defence compounds are general toxicants designed to dissuade any organism from consuming the plant. Natural does not necessarily equal healthy!

Azadirachtin (Figure 3.22) is an example of a compound that is becoming increasingly popular as a 'green' insecticide. It is extracted from the seeds of the neem tree, *Azadirachta indica*, and shows a multitude of biological effects on insects. (For a comprehensive review of the chemistry and biological activity of azadirachtin and related tetranortriterpenoids see Schmutterer.[11]) Azadirachtin is reported to show a wide range of effects on insects with different species showing different susceptibilities to the compound. It shows antifeedant and deterrent effects, clearly a useful property for the protection of treated crops, but there is evidence that in the case of 'no choice' the treated crop will be consumed, albeit at a slower rate than would have been expected. Insects consuming quantities of azadirachtin show a number of effects. Growth regulation is manifested in growth and moulting abnormalities

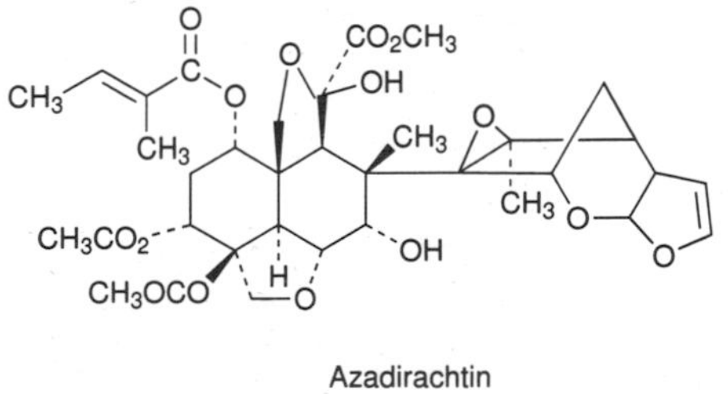

Azadirachtin

Figure 3.22 *Azadirachtin*

brought about by both a disruption of the endocrine system through blockage of release of neurosecretory peptides that regulate synthesis of ecdysone and juvenile hormone and through direct effects on dividing cells. [Ecdysone is the insect hormone that initiates the shedding of the larval skin (ecdysis) between instars]. Recent work by Mordue[12] has shown that azadirachtin binds to the testes of mature desert locusts, *Schistocerca gregaria*, and prevents the formation of sperm within the insect.

The neem tree comes from the plant family Meliaceae (the mahogany family), and several related species produce compounds with insecticidal properties. Notable amongst these are marrangin from *Azadirachta excelsa*, meliacarpins and meliatoxins from *Melia azadarach* and toosendanin from *Melia toosendan* (Figure 3.23). There is much less information on the effects of these compounds on insect species but it is believed that they act through a similar mode of action to azadirachtin, giving similar metamorphosis disturbing responses.

Ryanodine (Figure 3.24) is derived from *Ryania speciosa* and represents the first successful discovery of a natural insecticide in a collaborative programme between Merck and Rutgers University. The compound acts by binding to the calcium channels in the sarcoplastic reticulum of muscles, thereby causing calcium ions to enter the cells with death following rapidly thereafter.

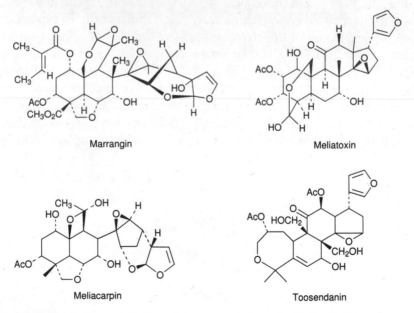

Figure 3.23 *Typical insecticidally active compounds from the family Meliaceae*

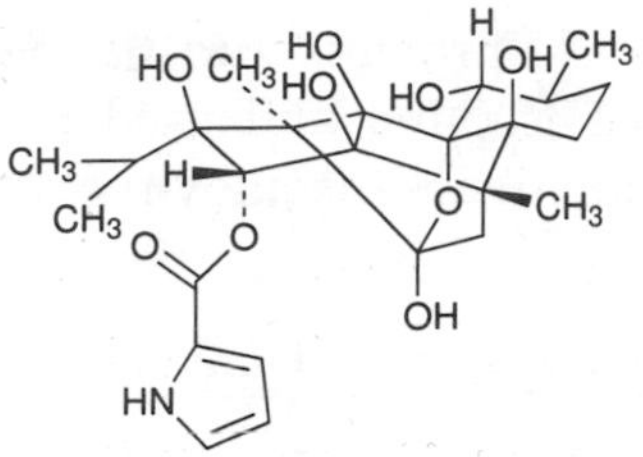

Figure 3.24 *Ryanodine*

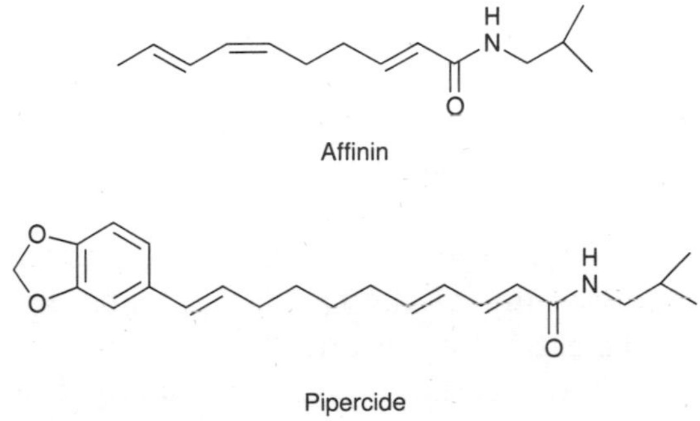

Affinin

Pipercide

Figure 3.25 *Affinin and pipercide*

There was some interest in the isobutylamides derived from various plant species including affinin (Figure 3.25) from *Heliopsis longipes* and pipercide (Figure 3.25) from *Piper nigrum*. These compounds are potent voltage-dependent blockers of the sodium channel but are too unstable to be used as products directly.

4.3 Insecticidal Compounds from Animals

Animals often use toxins to immobilise their prey, often insects. Wasps, bees, spiders, mites, scorpions, snakes and other reptiles are all capable of producing potent toxins many of which are insect specific. There is much work in progress around the world examining the opportunities that exist to exploit these toxins to produce new insecticides. This is usually undertaken in two different ways. The first is to determine the mode of action of the natural toxin and to use this novel effect to find synthetic compounds with insecticidal activity in biochemical screens. The second is to attempt to synthesise compounds with the same structural features of the natural toxin and hence with the same mode of action but with better stability following application. The types of compounds that are known are discussed by Blagbrough and Moya[13] but none has been commercialised to date.

Some of these toxins are proteins and the genes for several insect-specific toxins have been isolated and used to transform both crops (Transgenic Crops, Section 6) and baculoviruses (Section 5.1).

5 LIVING SYSTEMS FOR INSECT CONTROL

There is continual pressure from environmental groups to reduce the use of chemicals for the protection of plants. Insect and mite control can be achieved through the use of several different living systems although it must always be remembered that an insect pathogen requires time to enter the host and time to kill it. Good pathogens will kill their hosts slowly so as to ensure the opportunity to maximise their own reproduction and thereby propagate. In addition, if the pathogen has to enter the insect through the insect cuticle, conditions on the crop must be appropriate for the germination of the spore in the presence of the host insect and for the viability of the spore in its absence. Orally active compounds must be eaten by the target insect before they exert their effects. This will almost certainly be a relatively slow process and significant crop damage may occur before the pest succumbs. Glass-houses offer relatively uniform and protected environments and biological pest control systems work very effectively in them. Field crops, however, are much more variable in their growing conditions and pest pressure is also variable in terms of the intensity of attack and the pest species involved in the attack. This often presents too difficult a problem for a biological agent alone to be effective.

5.1 Baculoviruses

Viral diseases have been found in over 600 species of insect and the most important taxonomic group is the Baculoviridae because they are specific to invertebrates, are infectious by ingestion to insects that are important pests of agriculture and forestry and they exhibit excellent horizontal transmission. Baculoviruses contain three subgroups – sub-group A, the nuclear polyhedroviruses; sub-group B, the granuloviruses; and sub-group C, the non-occluded viruses.

Baculoviruses are generally named after the insect from which they were first isolated. For example, *Cydia pomonella* granulovirus (CpGV) was isolated initially from codling moth larvae and *Anagrapha falcifera* nuclear polyhedrovirus (AfNPV) was isolated from the alfalfa looper. This nomenclature is now widely accepted but it often conceals the number of insects that can be infected and the preferred insect host of the virus.

The problems with baculoviruses as insect control strategies are many and include the slow rate of kill, instability on crop foliage, expense of production and the specificity of spectrum. Farmers are used to rapid, broad-spectrum insect pest control and the response shown by viruses is generally unacceptable. However, there are situations where a crop is tolerant of some insect damage and is attacked by a single insect. Forestry is one such situation and control of a number of forest pests has been achieved using baculoviruses. These include gypsy moth (*Lymantria dispar*), Douglas fir tussock moth (*Orygia pseudotsugata*), pine beauty moth (*Panolis flammea*) and sawflies (*Neopridion* spp.).

Work is in progress to attempt to speed up the speed of kill of insects and to broaden the spectrum of biological activity. One way is to introduce genes that code for insect specific protein toxins into the baculovirus, often in place of the viral gene that codes for ecdysteroid UDP-glucosyltransferase (EGT), a protein that inhibits normal insect moulting in the infected insect, thereby causing the infected insect to continue feeding. Typical inserted genes code for insect-specific toxins from the scorpions *Buthus eupeus*, *Androctonus australis* and *Leirus quinque-striatus*, from the itch mite *Pyemotes tritici* and insect hormones such as diuretic hormone, juvenile hormone esterase and eclosion hormone. If a virus prevents feeding and kills its host more quickly, then the number of virions produced in the infected caterpillar will be lower than with a wild-type virus and the risk of the engineered virus outcompeting the wild type is removed. Preliminary trials by several companies have shown that this engineered baculovirus approach to insect control can give excellent results.

5.2 Bacteria

The key insect-specific bacterium is *Bacillus thuringiensis* and this is through the production of delta-endotoxins. It has been shown that if the endotoxins are applied to crops together with living spores of the bacterium, insect control is improved. It is common for living formulations to be sold.

5.3 Entomopathogens (Fungi)

A number of fungi are excellent insect pathogens. Some have been commercialised and others are under development and these find a place in insect control in covered or protected crops and in some situations for the control of insect swarms.

The main entomopathogens are *Verticillium lecanii* for control of aphids, white fly and thrips, *Beauveria bassiana* and *B. brongniartii* for control of lepidopteran and coleopteran species and *Metarhizium anisopliae* and *M. flavoviride* for control of insect pests of coffee and sugar cane and, more recently, to control locust swarms.

5.4 Nematodes

The nematode species of most interest for the control of crop insect pests are from the genera *Steinernema* and *Heterorhabditis*, both characterised by their association with bacteria from the genera *Xenorhabdus* and *Photorhabdus*. These nematodes invade the insect and then release their symbiotic bacteria into the insect's haemocoel. The nematodes release metabolites that repress the immune system of the insect, allowing the bacteria to develop. The bacteria then release toxins that kill the insect within two to three days and also produce antibiotics that prevent the invasion of the dead insect by other bacteria. These bacteria then invade the entire insect cadaver and the nematodes subsequently begin to consume the bacteria.

To date these nematodes have been shown to be effective at controlling soil-inhabiting insects but their commercial exploitation has been hampered by the cost of their production.

5.5 Insect Predators and Parasites

Many insects are the natural food source for other invertebrates. A number of these insect parasites and predators are used in some situations and give excellent insect control. Examples include the use of *Encarsia formosa* to control white fly (*Bemisia tabaci* and *Trialeurodes vaporariorum*) and *Amblyseius degerans* to control thrips in glasshouse crops and *Phytoseiulus persimilis* to control red spider mites in glasshouse and outdoor crops. There are many such invertebrate species sold and they have achieved wide acceptance in situations where the use of chemical insecticides is unacceptable. Typical situations include glasshouse crops pollinated through the use of bumble bees.

6 TRANSGENIC CROPS

It has already been pointed out that many naturally occurring insect-specific toxins are proteins. This means that the genes coding for those proteins can be isolated and introduced into crop plants, driven by selected promoters that allow the insecticidal proteins to be produced within the plant tissues. Typical genes that have been expressed in crop plants include truncated *Bt*-genes in cotton, maize and potatoes, and work is underway to evaluate the potential of genes coding for the snowdrop lectin and cowpea trypsin inhibitor as insect control strategies. Cotton transformed with *Bt*-genes is entering its second year of commercial sales and has been widely acclaimed by farmers as a significant contribution to the control of insects in the crop. Preliminary indications have shown that the use of transgenic cotton may reduce chemical insecticide inputs into the crop by as much as 60%. There are worries, however, that the presence of the toxin in the crop throughout the season may increase the onset of resistance. This is being addressed in many ways. One is to implement resistance control strategies whereby a proportion of a crop is not transgenic and is not treated with *Bt*-based insecticides in order to ensure that an area is available within which insect pests can feed without exposure to the toxins. A second strategy is to introduce two or more genes into the crop coding for different toxins with different modes of action. It is expected that crops transformed with two or more genes will be commercialised within the next two to three years.

7 RESISTANCE

Following the introduction and extensive use of chemical insecticides, a number of insects developed resistance to those insecticides. This meant that the usual dose that was effective in controlling the wild-type population no longer protected the crop and so farmers applied higher doses. This increase in the amount of compound applied introduced a higher degree of selection pressure on the insects that were resistant to the detriment of the susceptible organisms. Within a very short period, the use of some compounds was completely ineffective in controlling the phytophagous species at which they were targeted.

The mechanisms of resistance fall into two main categories. Many insects produce an increased level of detoxifying enzymes, such as esterases, that modify the insecticides to inactive metabolites very rapidly. Such a system is seen in aphids that are resistant to OP insecticides. In other cases it is the target site that is modified such that the insecticide (the enzyme inhibitor) no longer binds to the target and is, therefore, ineffective. This has recently been shown to occur in some aphids that are resistant to OP insecticides but the classical example is knockdown resistance (*kdr*) and *super-kdr* to pyrethroid insecticides shown by many insects but particularly house flies (*Musca domestica*). This resistance is thought to result from a modification of the voltage-sensitive sodium channel, the primary target site for these insecticides.

The introduction of transgenic crops that produce insect-active toxins throughout the life of the crop are suggested by some as being a strategy that is bound to be disastrous in terms of encouraging resistance to natural insecticides. *Bacillus thuringiensis* toxins are the primary candidates for incorporation into crops and it has been shown recently that diamondback moth larvae (*Plutella xylostella*) that have been treated repeatedly with *Bt*-based foliar sprays have developed resistance to the toxins. The opportunity, with the constant expression of the toxin in crops that are infested with insects, for resistance to develop is huge. Regulatory authorities are demanding that companies who introduce transgenic insect resistant crops also implement resistance management strategies to reduce the possibility of this occurring.

In addition, agrochemical companies continue to search for new compounds with novel chemistry and with new modes of action that can be incorporated into crop protection management strategies that will control resistant insects. Furthermore, there is also the opportunity to introduce synergists of insecticides that depress or inhibit the detoxifying enzymes and, thereby, enhance the effectiveness of the compounds.

The industry-led Insecticide Resistance Action committee (IRAC) records cases of insect resistance by crop and recommends strategies to reduce the impact and incidence of resistance.

8 QUESTIONS

3.1 Why is it likely that insecticides will be more toxic to non-target organisms such as beneficial insects and mammals (including man) than herbicides or fungicides?

3.2 Identify two biochemical processes that are not involved in nerve function but might be good targets for a new insecticide?

3.3 Identify four widely used natural insecticides and a major class of synthetic compounds that were derived from a natural insecticide?

3.4 Why is it preferred to have a compound that is effective through contact rather than as a stomach poison?

3.5 What are the advantages and disadvantages of introducing genes coding for insect-toxins into major crop plants?

3.6 Devise a strategy to avoid or delay the onset of insect resistance using chemicals (both insect toxic and behaviour modifying), predators and parasites and transgenic crops?

3.7 Is it acceptable to apply synthetic chemicals to crops to protect them from insect attack?

9 REFERENCES

1. Holy Bible, King James Version, Exodus, Ch. 8, verses 21–24; Ch. 10, verses 4–15, 1611.
2. D. Pimentel (ed.), 'Techniques for Reducing Pesticide Use: economic and environmental benefits', Wiley, Chichester, 1997.
3. D. Pimentel, 'Species Diversity and Insect Population Outbreaks', *Ann. Entomol. Soc. Am.*, 1961, **54**, 76–86.
4. J.A. Benson, 'Toxins and Receptors: leads and target sites', in Neurotox '91: Molecular Basis of Drug and Pesticide Action', ed. I.A. Duce, Elsevier Applied Science, London, 1991, pp 57–70.
5. C.D.S. Tomlin (ed.), 'The Pesticide Manual', 11th Edn., British Crop Protection Council, Farnham, UK, 1997.
6. I. Schmeltz, in 'Naturally Occurring Pesticides', eds. M. Jacobsen and D.G. Crosby, Marcel Dekker, New York, 99 pp.
7. A. Elbert, H. Overbeck, K. Iwaya and S. Tsuboi, 'Imidacloprid, a Novel Systemic Nitromethylene Analogue Insecticide for Crop Protection', in 'Brighton Crop Protection Conference – Pests and Diseases 1990', British Crop Protection Council, Farnham, UK, 1990, Vol. 1, pp. 21–28.
8. M. Elliott, 'Synthetic Insecticides Related to Natural Pyrethrins', in 'Crop Protection Agents from Nature: natural products and analogues', ed. L.G. Copping, Royal Society of Chemistry, Cambridge, UK, 1996, pp. 254–300.
9. F. Colliot, K.A. Kukorowski, D. Hawkins and D.A. Roberts, 'Fipronil: a new soil and foliar broad spectrum insecticide', in 'Brighton Crop Protection Conference Pests and Diseases, 1992', British Crop Protection Council, Farnham, UK, 1992, Vol. 1, pp. 29–34.
10. L.F. Adams, C.-L. Liu, S.C. MacIntosh and R.L. Starnes, 'Diversity and Biological Activity of *Bacillus thuringiensis*', in 'Crop Protection Agents from Nature: natural products and analogues', ed. L.G. Copping, Royal Society of Chemistry, Cambridge UK, 1996, pp. 360–389.
11. H. Schmutterer (ed.), 'The Neem Tree *Azadirachta indica* A. Juss. and Other Meliaceous Plants: Sources of unique natural products for integrated pest management, medicine, industry and other purposes', VCH, Weinheim, 1995, 696 pp.
12. A.J. Mordue, 'Actions of Azadirachtin, a Plant Allelochemical, against Insects', *Pesticide Sci.*, 1997 (in press).
13. I.S. Blagbrough and E. Moya, 'Animal Venoms and Insect Toxins as Lead Compounds for Agrochemicals', in 'Crop Protection Agents from Nature: natural products and analogues', ed. L.G. Copping, Royal Society of Chemistry, Cambridge, UK, 1996, pp. 339–359.

CHAPTER 4

Fungicides

1 INTRODUCTION

Although they share some characteristics with animals and plants, the fungi are phyllogenetically distinct. Fungi are nucleated and possess well-defined cell walls but they are spore-bearing and do not photosynthesise. They reproduce asexually or sexually and usually exhibit filamentous growth but some have life cycles that are characterised by the inclusion of a motile, free swimming phase.

Over 200 000 species of fungi have been characterised. The fungi are extremely responsive to environmental pressures and exhibit a high capacity to adapt to and colonise a variety of ecological niches. They can be found growing anywhere between the high mountains to the abyssal depths and from dry, polar regions to the hot, humid tropics.

The absence of chlorophyll restricts fungi to either a saprophytic or parasitic existence, or a combination of both, depending upon the environment and the availability of substrate. As saprophytes, fungi are responsible for much of the degradation of organic matter, release of plant nutrients and their recycling in natural and agronomic systems. They play an essential role as fermentation agents in the production of antibiotics, bread, beer, wines and in the manufacture of cheese and a range of secondary metabolites, including some important fungicides. The same saprophytic life-style also means that certain fungi destroy food, clothing and other manufactured articles that have an organic base. As pathogens, fungi can infect a wide spectrum of animals, including man, insects and nematodes, and are the main cause of plant disease. Although most plants are immune to most fungal diseases, each plant species has an attendant following of pathogens. For example, rice is attacked by over 20 fungi, each capable of causing severe economic damage.

Plant pathogenic fungi have played a significant part in shaping human history.[1] Their effects are recorded in the Bible as 'blasting and mildew' (Deuteronomy 28, verse 22; Amos 4, verse 9) and the Greeks and

Romans were acutely aware of their potential damaging nature and of the fungicidal activity of sulfur. The threat of crop losses due to fungal attack even prompted the Romans to make sacrifices to their deities in an attempt to control diseases of cereals. Judging by the absence of this as an accepted method of disease control this is probably one of the first records of an ineffective fungicide, or at least of its incorrect application. Better known examples of the historic importance of plant pathogens are 'St Anthony's Fire' and potato blight. The former, caused by the ergot fungus *Claviceps purpurea*, was common in Europe in the Middle Ages and was characterised by severe haemorrhage, abortion and death of those unfortunate enough to eat contaminated rye bread. Epidemics of potato blight in Europe in the late 19th century led to widespread famine and, in part, to the emigration of 1½ million people from Ireland to the New World, ironically from where the cause of the blight, *Phytophthora infestans*, had originated. The use of monoculture encourages fungal epidemics and the dangers that presented themselves to farmers over 100 years ago are still with them, controlled only by the careful and directed use of fungicides and modern husbandry techniques.

The agriculture and horticulture of France has always had a crucial role in the development of fungicides, beginning with the prize, offered by the French Academy of Arts and Sciences in 1750, for the most informative work on the cause and control of wheat bunt, *Tilletia caries*. It was won by Tillet and pioneered the use of fungicidal seed treatments containing such diverse materials as lime, salt, potassium nitrate and urine. Subsequent importations of infected plant material into Europe from the New World threatened the French vine industry on three separate occasions and forced French plant pathology to the forefront of agricultural research.

The first was the rapid spread of *Uncinula necator*, vine powdery mildew, following its identification in England in 1845. A search for control methods led, from the initial observation by Mr Tuker in England that sulfur was an effective treatment, to the development in 1855 by Bequerel of a fungicide application programme based on the use of a fine form of sulfur dust.

The second invasion came in 1865 when attempts to control the root aphid *Phylloxera* through the importation and use of *Phylloxera*-resistant root stocks inadvertently introduced vine downy mildew, *Plasmopara viticola*, into the French vineyards. This stimulated what is regarded as the major breakthrough in fungicide use when Millardet, in 1885, acting upon a chance observation, developed the use of Bordeaux mixture, a cocktail of copper sulfate and lime, which is still used extensively in vines and many other crops.

The advances in disease control made in Europe and the associated increases in crop yields prompted further research in the US. Whilst the ensuing collaborative efforts of the French and American pathologists undoubtedly impacted upon the development of, in particular, copper and sulfur fungicides, such opportunities for travel between continents are never missed by Nature and it is likely that the appearance in France at that time of *Guignardia bidwellii*, another pathogen of vines native to the New World, was no coincidence.

The use of sulfur and copper fungicides dominated the market until 1915 when organomercury seed treatments, derived from discoveries in the pharmaceutical and dyestuffs industries, were shown to be highly effective against cereal diseases. They remained popular until banned in the late 1970s on the grounds of adverse toxicology.

Up to the mid-1960s, the fungicide market comprised broad-spectrum, high rate of use, multi-site of action, non-systemic protectants (see later). The announcement of systemic compounds caused a revolution in farmer practice, permitting much greater flexibility of use with increased levels of control. New opportunities for fungicides were recognised, particularly in major European crops. Intensive cereal production, characterised by high density monoculture and high levels of fertiliser use, was especially responsive and large yield increases were recorded following the adoption of the new systemics for the control of the 'ancient' diseases caused by powdery mildew and rust.

2 FUNGICIDE PERFORMANCE

Growers use fungicides in combination with other crop management products and practices to increase profit through their effects on qualitative or quantitative crop yield. In common with all other crop protection products, the commercial success of fungicides depends on their biology and not upon their chemistry or mode of action. Growers are interested only in product performance and how it affects their overall ability to control disease to an acceptable level. The measure of performance includes important economic features such as the restrictions on timing and frequency of fungicide application, compatibility with other products, cost and safety.

In particular, growers are concerned about the risk of resistance, which first emerged as a serious problem in the 1970s following the introduction and extensive use of systemic fungicides. The problem affects the grower and fungicide manufacturer because of the loss in profit. Potentially, it affects consumers because of the eventual reduction in food variety and quality, and increased prices. It is the ever present risk of resistance development, as much as the need to increase efficacy, that drives the fungicide discovery effort. Anti-resistance strategies involve the use of product mixtures or alternating application of compounds with a different mode of action.

Three phases of fungicide performance can be distinguished. The earliest, up to the 1960s, saw the development of broad-spectrum, multiple site-of-action compounds. Characteristically, these were non-systemic and were used solely as protectant materials. Many are still important products and are especially useful in the control of fungi resistant to the more specific acting fungicides. The discovery of single site-of-action fungicides heralded a revolution in fungicide performance. Their specificity not only allowed new systemic activity to be exploited, and hence curative control, but also enhanced efficacy levels and long-term control. In some crops, such as cereals, the systemic fungicides provided the farmer with the first practical means to control economic-ally important diseases. Their use expanded rapidly through the devel-opment of new markets and by the replacement of comparatively inflexible protectant fungicides. However, it appears likely that their success and, indeed, the success of all site-specific fungicides, will always be threatened by the capacity of fungi to evolve resistant strains. Paradoxically, the development of systemic fungicides is founded upon the continued stability of the non-systemic market.

As we enter a new millennium, fungicide technology is on the thresh-old of a third phase of change. Materials are now available that lack intrinsic fungicidal action but are effective in 'switching on' the inherent resistance mechanisms of plants. Their exact mode of action is a subject for debate. On the one hand, plants are subject to continual challenges by fungi, most of which fail. It seems reasonable, therefore, to postulate that natural resistance mechanisms are 'switched on' at an equal rate. How is it then that an exogenously applied compound that stimulates the plant's defence systems into action can operate to prevent disease in situations that are already providing a similar stimulus? However, it is claimed that a bright future exists for this new generation of disease control agents although ultimately their success will rest heavily on the ease at which resistance to the induced defence mechanisms can develop.

3 PRODUCT ATTRIBUTES

Fungicides are either mobile or immobile in plants.

Systemic fungicides are mobile compounds that are able to penetrate the cuticle of leaves and stems and enter the plant symplast or apoplast. These are generally single site of action compounds, very effective at low rates of application, flexible to use (that is, with acceptable efficacy over a wide range of fungal growth and infection stages) and provide disease control for extended periods. Few fungicides possess the physico-chemical structure to enable them to penetrate the symplast and move within the phloem-mediated transport system. Most fungicides show a degree of basipetal movement, implying symplastic transport, but this is usually caused by simple diffusion and/or vapour phase activity. In general, however, fungicides move only within the apoplast, demonstrating translaminar mobility or redistribution throughout the plant via the transpiration stream. Thus, most fungicides, when applied to leaves, move towards the leaf margins.

An important factor in the redistribution of fungicides throughout a target crop is their level of vapour phase activity. Compounds that are relatively immobile in the plant may be extensively redistributed through the vapour phase and effect commercially acceptable disease control. A good example is fenpropimorph.

Immobile fungicides reside entirely at the site of application. Although their redistribution within a crop can be modified by the use of adjuvants or by the use of spraying techniques designed to increase the penetration of spray into the foliage, they cannot move once applied other than through the action of rain. This, however, is unlikely since movement in rainwater implies a degree of loss of fungicide and is usually prevented as far as possible by formulation. Because they are surface-bound compounds, immobile fungicides can be multi-site inhibitors and have a corresponding broad spectrum of activity. In growing crops, the expansion of leaf, stem and fruit surface tissue breaks the protective layer formed by the fungicide, allowing fungal pathogens to invade. Consequently, immobile fungicides are applied frequently. In vines, for example, applications may be made at 10 day intervals throughout the season. The multi-site nature of their activity is at once a weakness, in the limitation of their use to surface fungicides, and a strength, in their use in anti-resistance strategies.

Fungicides are eradicants and/or protectants.

Disease control may be effected following symptom expression, as in the eradication of visible powdery mildew colonies on wheat, or prior to disease expression. The difference is embraced in the use of the terms eradicant, curative and protectant. Strictly, *eradicant fungicides* are active only against the later, visible stages of the fungal life cycle, for example, the external mycelium or spore-bearing structures formed after fungal establishment in the host, or structures internal to the plant that cause changes in the appearance of the infected tissue, for example, chlorosis. *Curative fungicides* are active against the early but post penetrative events in the infection process. The visible effect of a curative fungicide is the same as a protectant compound but the flexibility, that is, the period during which control can be effected, is much greater with a curative and quickly becomes evident in practice. *Protectant fungicides* prevent infection. They are active against spore germination, germ tube development and growth and appressorium formation.

Traditionally, protectants were synonymous with immobile fungicides. Some, however, demonstrate remarkable mobility. Quinoxyfen is a new fungicide, active against members of the Erysiphales in temperate cereals, vines and fruit. Disease control is achieved through the inhibition of appressorium formation, a process that usually occurs within a few hours from the impact of the conidium on the host leaf surface. The compound has been shown to be effective against mildew even on leaves not present at the time of treatment, implying its movement through the apoplast or symplast, or both, or via the vapour phase to the potential infection site. The strobilurins are inhibitors of complex III in respiration and show broad-spectrum disease control. Although they are known to penetrate plant tissue, their major fungicidal activity arises from the inhibition of fungal growth stages that occur prior to penetration. Their success as protectant fungicides, therefore, is reliant upon their slow redistribution in treated crops, a commercially significant advantage over most other surface active products.

Some compounds, marketed as fungicidal, act indirectly. Two groups, the pro-fungicides and the so-called 'plant activators' are important. In each group, direct fungicidal activity is absent or minimal. *Pro-fungicides* are compounds that when applied to target crops are metabolised into fungicidal products, either by the crop or the fungus. Theoretically, they can be used to respond to the different physico-chemical characteristics required for penetration, movement in the plant and uptake by the fungus. *Plant activators* operate by triggering the systemic acquired resistance (SAR) mechanism in host plants. Although SAR has been known for many years, CGA 245704 (acibenzolar), which was announced by Ciba in 1994, is the first to attempt to exploit it as a means of disease control.

4 MODE OF ACTION AND SELECTIVITY

The fungi are classified broadly into the lower fungi, collectively known as Phycomycetes, and the higher fungi, described by the classes Ascomycetes, Deuteromycetes and Basidiomycetes. The classification is indistinct; some taxonomists argue that the Phycomycetes are not true fungi.

The Oomycetes, a class within the Phycomycetes, is unusual and regarded by many as being distinct from the true fungi. The taxonomic argument to support that conviction is beyond the scope of this chapter but the differences that it describes are important in understanding fungicide selectivity. For example, the absence of chitin in the Oomycete cell wall and its presence in other fungi is a critical feature in the mode of action of several fungicides.

Fungicides are selective through their placement on the target crop, because they inhibit biochemical pathways that are absent or less sensitive to the applied compound in higher plants or because they are metabolically fragile in less sensitive organisms.

Several biochemical processes have been validated as targets for fungicide activity. They include the direct and precise inhibition of biosynthetic sequences:

- Sterol biosynthesis;
- glycerophospholipid biosynthesis;
- nucleic acid metabolism;
- tubulin biosynthesis;
- chitin biosynthesis;
- melanin biosynthesis;
- protein biosynthesis;
- respiration inhibitors.

or have been identified as having a general disruptive effect, for example, through their interference with cell membrane integrity or by the non-selective destruction of enzyme function.

However, the oldest and still the most commercially valuable group of fungicides are the general cell toxicants.

4.1 General Cell Toxicants

This group comprises a chemically diverse range of compounds, including heavy metals, inorganic compounds and organic complexes (Figure 4.1). Although some have been implicated as inhibitors of respiration and have also been shown to precipitate protein at high concentration, other more general action is likely, including binding to enzymes containing sulfhydryl or amino groups, resulting in their inactivation and a general loss of cell integrity.

4.1.1 Inorganics. Sulfur was the first effective fungicide and is still widely used as a protectant acting through the inhibition of conidial germination. The selective activity of sulfur against powdery mildews

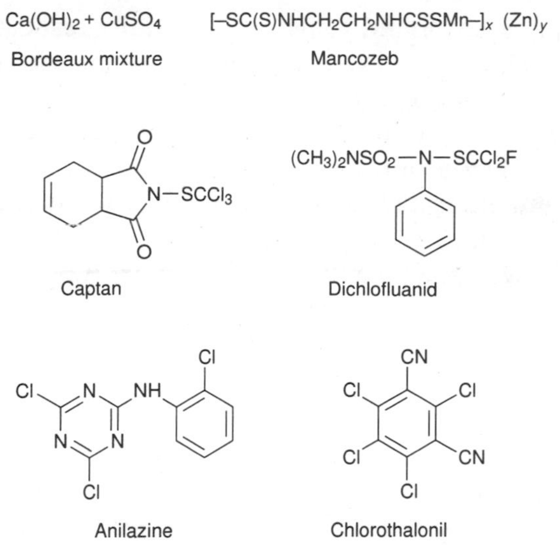

Ca(OH)$_2$ + CuSO$_4$ [–SC(S)NHCH$_2$CH$_2$NHCSSMn–]$_x$ (Zn)$_y$

Bordeaux mixture   Mancozeb

Captan

(CH$_3$)$_2$NSO$_2$—N—SCCl$_2$F

Dichlofluanid

Anilazine

Chlorothalonil

Figure 4.1 *General cell toxicants*

may be attributed to their unique and exposed growth habit or to possible uptake by the lipid layers of conidia.

Copper, as copper sulfate, was first used to control *Tilletia caries* in wheat but its main use, as Bordeaux mixture, is in the control of Oomycetes in a wide spectrum of crops such as potato, tomato and vines or in combination with systemics such as cymoxanil.

In the 1930s, the use of organomercurial fungicides became popular primarily as seed treatments but it has since been banned because of its high toxicity and persistence in the environment.

4.1.2 Organics. Organic multi-site inhibitors include two major groups, dithiocarbamates and phthalimides. Both groups display a broad antifungal spectrum and are applied as foliar, soil and seed treatments in fruit, vine and vegetables (especially potato). However, the dithiocarbamates and phthalimides lack activity against the Erysiphales (powdery mildews). Examples of the dithiocarbamates are maneb, mancozeb, ziram, zineb, ferbam and thiram. In ziram, ferbam and thiram, the anion is the basis for activity mediated through complexes with enzymes containing Cu^{2+} or sulfhydryl groups, leading to their inactivation. Anion/Cu^{2+} complexes are thought to interfere with respiration and the anion itself also binds to redox compounds such as glutathione and cysteine. Zineb, maneb and mancozeb exert their action through their conversion into ethylene diisothiocyanate and subsequent reaction with enzyme sulfhydryl groups.

Captan, captafol and folpet preferentially react with enzyme sulfhydryl groups but may also attack amino groups and inhibit enzymes that do not contain sulfhydryl groups.

Other multi-site of action organic fungicides include dichlorofluanid, anilazine and chlorothalonil.

Dichlofluanid is a close relative of the phthalimides and has the same mode of action. Replacement of the chlorine atom by fluorine introduces some activity against the Erysiphales but has little effect on the major use spectrum.

Anilazine is a 1,3,5-triazine used as a foliar fungicide to control a broad range of fungal pathogens in vegetables, turf, wheat, coffee and ornamentals.

Chlorothalonil, a chlorophenyl introduced in the mid-1960s, is a major protectant fungicide. It is used alone or in mixtures to control *Septoria* spp. in cereals, *Phytophthora infestans* in potatoes and *Botrytis* spp. in vegetables.

4.2 Sterol Biosynthesis Inhibitors

Sterols are derivatives of terpenes, within the larger chemical family of terpenoids. Such compounds are widespread and numerous in eukaryotes. Although in many cases their function is not clear, others are known to have an essential role in growth, development, metabolism and the integrity of cells. Sterols are localised in cell membranes, conferring stability and controlling permeability.

In fungi, ergosterol, synthesised from acetyl-CoA, is the major sterol (Figure 4.2; Mercer[2]) and has an essential role in the maintenance of membrane function such that a reduction in its availability disrupts membrane integrity.

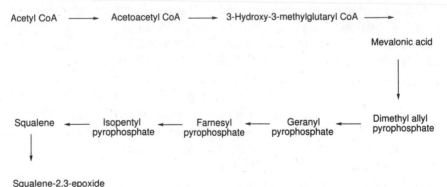

Figure 4.2 *Sterol biosynthetic pathway. Dotted arrows indicate several steps (continued opposite)*

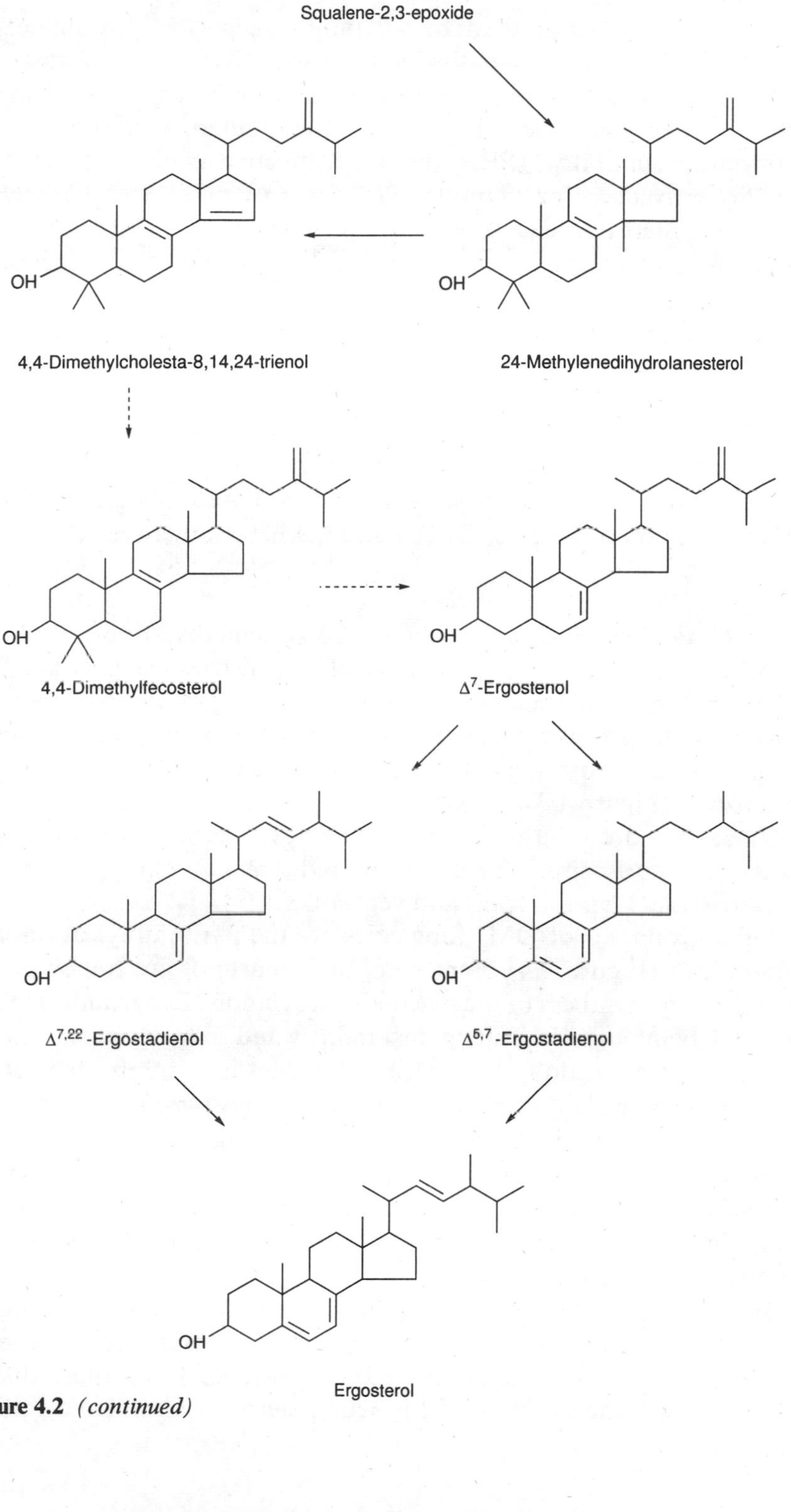

Squalene-2,3-epoxide

4,4-Dimethylcholesta-8,14,24-trienol

24-Methylenedihydrolanesterol

4,4-Dimethylfecosterol

Δ^7-Ergostenol

$\Delta^{7,22}$-Ergostadienol

$\Delta^{5,7}$-Ergostadienol

Ergosterol

Figure 4.2 *(continued)*

Ergosterol is synthesised by all fungi except the Phycomycetes in which the biosynthetic pathway is absent. Members of this group scavenge their sterol requirements from their hosts through mycelial uptake. This characteristic explains the limited spectrum of sterol biosynthesis inhibitors (SBIs) that are ineffective against diseases caused by Phycomycetes, for example, *Plasmopara viticola* and *Phytophthora infestans*. Sterol biosynthesis inhibitors are also unsuccessful in the prevention of spore germination, which is independent of ergosterol biosynthesis.

However, most commercially important fungal diseases are caused by Ascomycetes and Deuteromycetes and compounds that inhibit sterol biosynthesis are valuable crop disease control agents. In general, they are mobile within individual plants and, in some cases, throughout the crop via vapour phase activity, providing curative, eradicant and some protectant control. The SBIs are divided into the C14α-sterol inhibitors (demethylation inhibitors; DMIs) and the $\Delta^{8,7}$ isomerase, Δ^{14} reductase inhibitors.

4.2.1 C14α-Demethylation. The C14α-demethylation inhibitors (DMIs) comprise a diverse group of chemistries each containing a nitrogen-containing heterocycle attached to a lipophilic group in the 1 position. The group includes the most commercially valuable and most numerous family of fungicides, the azoles, comprising 1,2,4-triazoles and imidazoles (Figure 4.3).

The azoles are used to control a wide range of diseases, especially in cereals (*Septoria* spp., *Puccinia* spp. and *Pseudocercosporella herpotri-coides*), fruit (*Venturia* spp.) and vegetables.

Other examples of DMI fungicides are the pyrimidinylcarbinols and piperazines (Figure 4.4). Fenarimol and nuarimol are minor products and the only members of the pyrimidinylcarbinols. Fenarimol is the most versatile being used in fruit against mildew and scab diseases.

The principle action of the DMI fungicides is to inhibit the action of the cytochrome *P450*-dependent removal of the 14α-methyl group from 24-methylene dihydrolanosterol. The resulting accumulation of the substrate 24-methylene dihydrolanosterol, obtusifoliol and 14α-methyl-fecosterol, together with the consequent reduction in ergosterol synthesis, disrupts the normal functioning of cell membranes and was thought to be the basis of DMI activity.

However, recent studies suggest that the toxic activity resulting from the inhibition of C14-demethylation is more complex and the precise consequences are still uncertain.[3] The growth of some fungi does not correlate with the amount of 14-methyl sterols accumulated following

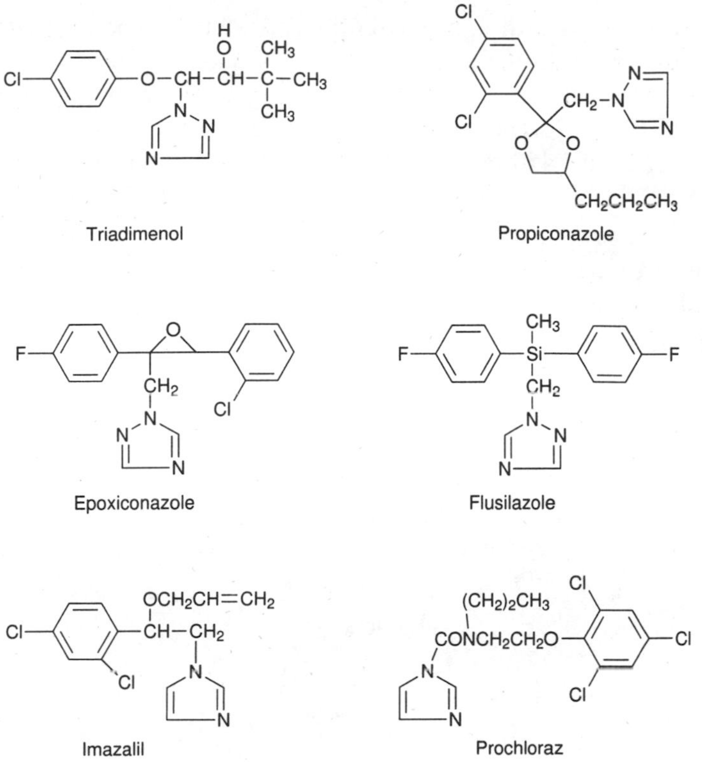

Figure 4.3 *Examples of 1,2,4-triazoles and imidazoles*

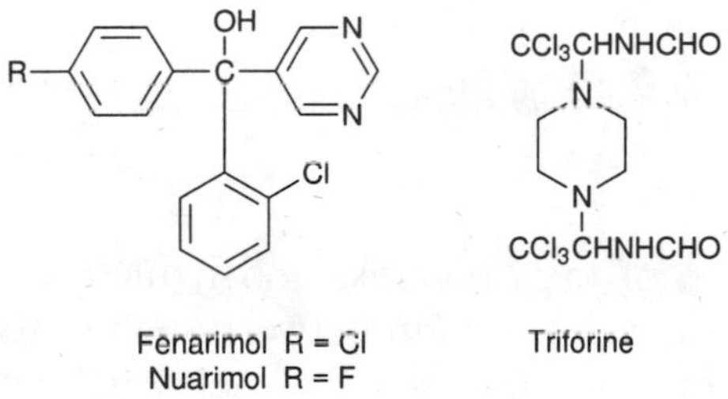

Figure 4.4 *Pyrimidinylcarbinols and a piperazine*

treatment with DMI fungicides. Although the primary action of DMIs on the demethylation step is not in doubt, the action of the resulting accumulation of precursor sterols may be questioned. Fungal mutants with defective C14-demethylase will grow if they also contain a defective $\Delta^{5,6}$ desaturase. Where the fungus possesses an active $\Delta^{5,6}$ desaturase, the bulk of the accumulated sterol precursor fraction is 14α-methylergosta-

8,24(28)-dien-3β,6α-diol, a sterol now considered to be the main agent in membrane disruption.

There are significant differences in performance between DMIs. These may be due to differences in their binding affinities to the haem moiety of the cytochrome *P*450-dependent demethylase enzyme.

4.2.2 $\Delta^{8,7}$ Isomerase and Δ^{14} Reductase. Before the appearance of resistance to azole fungicides in cereal powdery mildew the $\Delta^{8,7}$ isomerase and Δ^{14} reductase inhibitors were of minor importance because of their limited spectrum. Only seven examples of this group have shown commercial potential: fenpropimorph, fenpropidin, tridemorph, dodemorph, aldimorph, piperalin, the first member of the series, introduced by Eli Lilley (now Dow Agrosciences) in 1960, and spiroxamine, the latest addition (Figure 4.5).

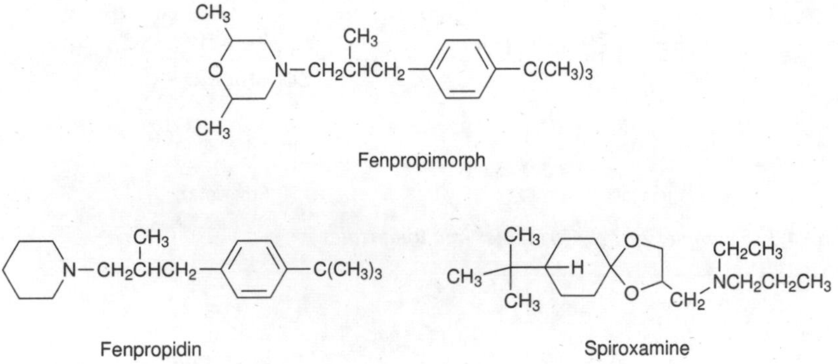

Figure 4.5 *Morpholines and piperidines*

The inhibition of the isomerase and reductase steps (Baloch et al.[4]; Köller[5]) involves an interaction between the negatively charged moiety of either enzyme and the positively charged nitrogen atom in the morpholine or piperidine ring of the fungicide. However, the balance of importance between the two enzyme steps is not well understood. Thus, it is known that whilst tridemorph strongly inhibits the isomerase step in the powdery mildew *Erysiphe graminis*, fenpropidin inhibits the reductase step and possesses only weak activity against the isomerase. Furthermore, fenpropimorph inhibits both enzymes. Additionally, morpholine inhibition of $\Delta^{24(28)}$ reductase, Δ^{24} transmethylation and squalene cyclisation steps may be possible modes of action.

The field performance of the morpholines and piperidines probably reflects a composite activity against all the possible biochemical sites since the fungicides are recommended to be used at high rates, for example, 750 gai/ha for fenpropimorph. Therefore, this group may be more strictly defined as comprising multi-site, systemic inhibitors, although the sites of action are confined to a single biochemical pathway.

4.3 Glycerophospholipid Biosynthesis Inhibitors

The glycerophospholipids form important complex derivatives in cells of all eukaryotes. They are essential to the functional integrity of membranes, in which, because of their anionic properties, they play a role in the regulation of cation transport. They also provide a permeability barrier to the movement macromolecules, and serve as a medium for the activity of membrane-associated proteins.

Two major pathways of phospholipid biosynthesis are recognised. These are the Bremner–Greenberg methylation pathway and the Kennedy cytidine nucleotide pathway (Figure 4.6). The basis for fungicidal selectivity in compounds that inhibit glycerophospholipid biosynthesis arises from the distinction between fungal systems of biosynthesis versus plant and animal systems. In fungi, the production of glyccrophospholipid is predominantly via the Bremner–Greenberg pathway, whereas in animals and plants the Kennedy pathway is the major route.

There are few commercial fungicides that have glycerophospholipid biosynthesis inhibition as their mode of action. The validated targets are phosphatidylcholine synthesis and phosphatidylinositol synthesis.

4.3.1 Phosphatidylcholine Synthesis. Phosphatidylcholine is essential for fungal cell growth. Inhibition of phosphatidylcholine results in vital changes in hyphal morphology such as decreased hyphal extension rates, increased branching, hyphal swelling and rupture. This activity is thought to be mediated by an associated reduction in chitin synthase activity.

The first phosphatidylcholine inhibitors were amongst the earliest modern commercial fungicides to be discovered. At the time of discovery, the mode of action was unknown.

Edifenphos, *O*-ethyl *S,S*-diphenyl phosphorodithioate, was announced in 1968 and introduced commercially by Bayer AG for the control of rice blast, *Pyricularia oryzae*. Iprobenfos, *S*-benzyl *O,O*-di-isopropyl phosphorothioate, was introduced by Kumiai Chemical

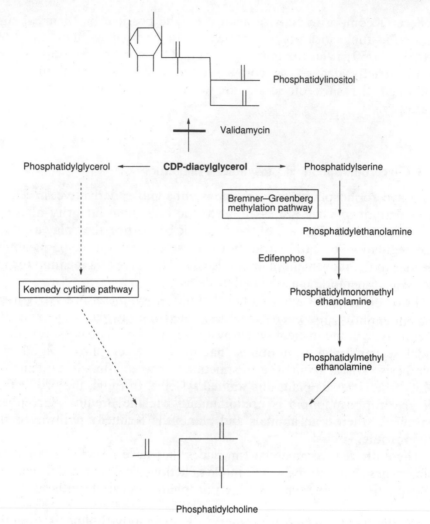

Figure 4.6 *Glycerophospholipid biosynthesis and fungicide mode of action (solid bars indicate sites of inhibition)*
(After Robson and Trinci, Ref. 6)

Industry Co. Ltd. Iprobenfos and edifenphos are used extensively in rice to control *P. oryzae*, *Corticium sasakii* and *Cochliobolus miyabeanus*.

Both compounds (Figure 4.7) are non-competitive inhibitors of chitin synthase and also interfere with the Bremner–Greenberg methylation pathway.[6] Their activity is dependent upon the oxidation of the phosphorous – sulfur linkage, a feature that is the basis of resistance development, which is common; sensitive strains of *P. oryzae* metabolise the compounds more rapidly than resistant strains.

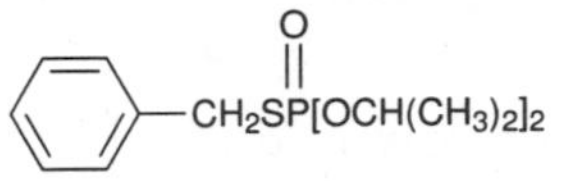

Figure 4.7 *Glycerophospholipid inhibitors*

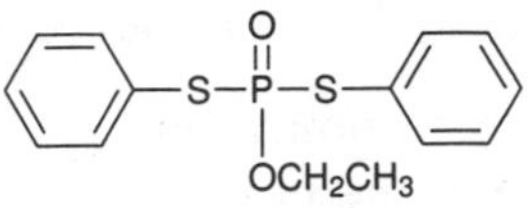

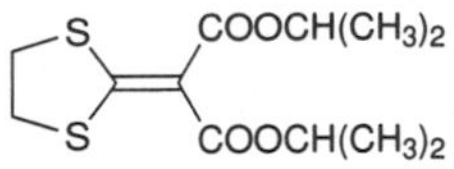

Figure 4.8 *Isoprothiolane*

Subsequent inhibitors of the Bremner–Greenberg pathway have been associated with high mammalian toxicity and no new representatives of the group have been launched commercially.

However, isoprothiolane (Figure 4.8) is cross-resistant to iprobenfos and edifenphos, suggesting a common mode of action.

4.3.2 Phosphatidylinositol Synthesis. The symptoms of inositol depletion include a decreased hyphal extension rate, increased branching and cell lysis. Other changes occur in protein, carbohydrate and nucleic acid metabolism, and reduced activity levels of membrane-bound wall biosynthetic enzymes.[7]

Validamycin A is the only commercial fungicide active against this target (Figure 4.9). The compound is an aminoglycosidic pseudosaccharide and is a secondary metabolite produced in the fermentation of *Streptomyces hygroscopicus* var. *limoneus*. Validamycin A was introduced by Takeda Chemical Industries Ltd. for the control of Basidiomycetes and Corticiaceae, especially sheath blight, *Rhizoctonia solani*, in rice.

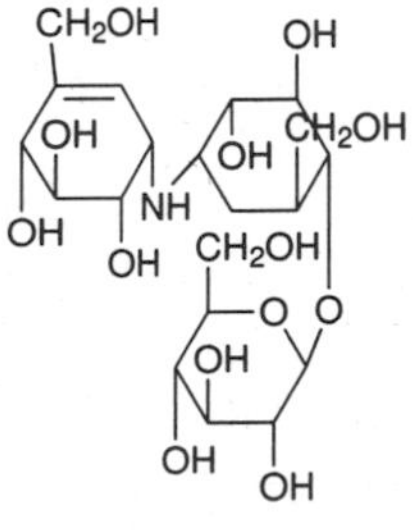

Figure 4.9 *Validamycin A*

4.4 Nucleic Acid Biosynthesis Inhibition

Nucleic acid metabolism is a feature that is common to all living things and few steps in the process in fungi are sufficiently different from those in other organisms to be used as an effective target for selective fungicides. Consequently, few compounds have been discovered that are not limited by their lack of specificity. In those fungicides known to inhibit nucleic acid biosynthesis the basis for the observed specificity is not understood.

Several compounds may interfere with nucleic acid metabolism but commonly their effects are secondary to their primary mode of action, for example, the benzimidazoles. Compounds that inhibit nucleic acid biosynthesis directly are either phenylamides, pyrimidines or hydroxy-pyrimidines. Recently, the phenoxyquinolines were identified as exhibiting a novel mode of action in purine biosynthesis and are potentially useful fungicides.

4.4.1 DNA Synthesis. Hymexazol is the only fungicide that has an implicated mode of action against DNA synthesis. It has broad spectrum activity and is used commercially in several crops, including rice and vegetables.

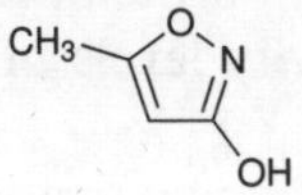

Figure 4.10 *Hymexazol*

4.4.2 RNA Synthesis. The phenylamides comprise a diverse group of systemic compounds of which the acylalanines are the most successful (Figure 4.11). The acylalanine fungicides are systemic, mainly via the

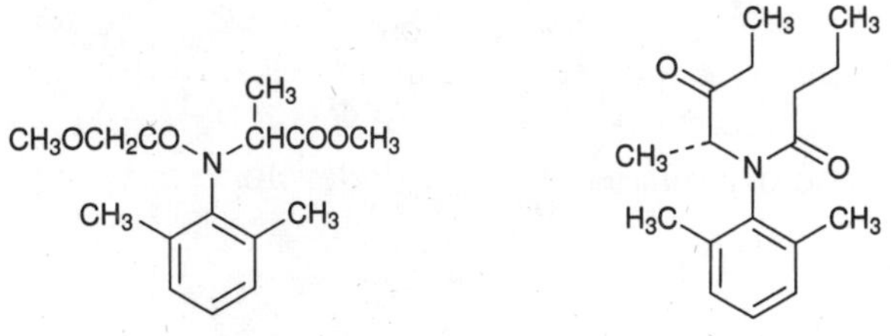

Metalaxyl

Metalaxyl M (CGA329351)

Figure 4.11 *Acylalanines*

transpiration stream, but some symplastic movement has also been reported. They are used commercially as protectants and curatives in seed treatments, root and foliar applications.

The acylalanines are characterised by metalaxyl, the most studied member of the group. Metalaxyl exists as two enantiomers, metalaxyl M being the most active. Metalaxyl is known to interact with the RNA polymerase–I–template complex,[8] inhibiting the incorporation of ribonucleotide triphosphates into ribosomal RNA.

The activity of the acylalanines is specific to the Oomycetes, in particular to the post-infection development stages. The reason for the taxonomic specificity is unknown but the activity against post-infectional stages is thought to reflect the dependence of organised hyphal growth upon the *de novo* synthesis of ribosomal RNA. The effects of metalaxyl upon the pre-infection stages, including zoospore formation, germination of zoospores or conidia, penetration and the formation of primary haustoria, are masked by a non-limiting supply of ribosomal RNA. Subsequently, as the ribosomal buffer is depleted, the inhibition of the RNA-polymerase I complex becomes effective and results in an accumulation of r-RNA precursors. Nucleoside triphosphates which promote fungal β (1,3)-glucan synthetase and stimulate the synthesis of cell wall constituents cause a characteristic thickening of hyphal walls.

The hydroxypyrimidines ethirimol, dimethirimol and bupirimate (Figure 4.12) are used specifically to control the powdery mildew diseases (Erysiphaceae).

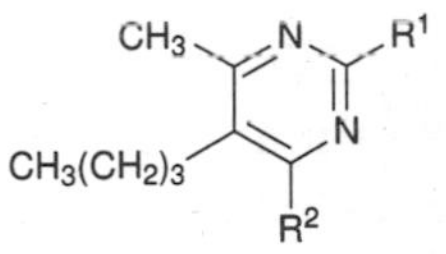

	R^1	R^2
Ethirimol	$NHCN_2CH_3$	OH
Dimethirimol	$N(CH_3)_2$	OH
Bupirimate	$NHCH_2CH_3$	$OSO_2N(CH_3)_2$

Figure 4.12 *Hydroxypyrimidines*

Hydroxypyrimidines inhibit adenosine deaminase, a key enzyme in the purine salvage pathway which recycles bases released during the breakdown of nucleic acids. The inhibition results in abnormal growth, particularly in those organisms or at those development stages in which *de novo* synthesis of purines is absent. Typically, hydroxypyrimidines inhibit the early development stages of powdery mildew infection, germ tube elongation and appressorium formation.

The absence of adenosine deaminase in plants is the reason for the crop selectivity of the hydroxypyrimidines. However, the fungicidal specificity of this group cannot be explained entirely on the basis of the distribution of adenosine deaminase, which occurs in a wide range of fungi. It has been suggested that since powdery mildews are unable to synthesise purines *de novo*, adenosine deaminase may be able to synthesise adenine and guanine nucleotides using purines scavenged from the plant host. However, the addition of inosine and other metabolites is ineffective in inhibiting the activity of ethirimol, hence it is unlikely that this mechanism has a significant function. Interestingly, only adenosine deaminase from the Erysiphaceae is sensitive to the hydroxypyrimidines.

The phenoxyquinolines are a recent addition to crop fungicides and only one compound has been described. LY214352, an experimental fungicide, inhibits a novel target, dihydro-orotate dehydrogenase (DHO-DH) in the pyrimidine biosynthesis pathway that catalyses the conversion of dihydro-orotate into orotic acid[9] (Figure 4.13).

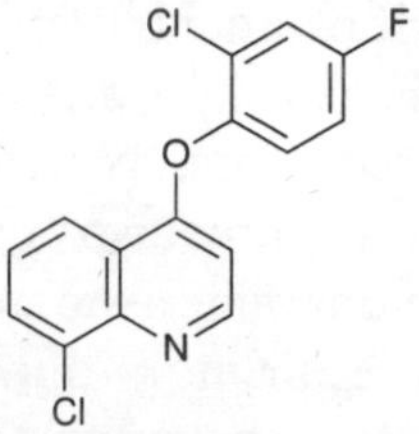

Figure 4.13 *LY214352 (phenoxyquinoline)*

4.5 Tubulin Biosynthesis Inhibition

One of the major features in the sequence of cell division is the formation of the mitotic spindle and the subsequent separation of chromosomes into their respective daughter cells. An important element of the spindle is the highly conserved, helical molecule tubulin. In addition to spindle formation and the segregation of chromosomes in cell division, alternating helices of α- and β-tubulin form the microtubules that form part of the cytoskeleton and have active roles in cell organelle organisation.

Benzimidazoles have a high affinity for tubulin proteins[10] and disrupt mitosis at metaphase by attacking the mitotic spindle. The ensuing failure of sister nuclei to separate results in cell death.

Benzimidazoles are highly selective. Oomycete fungi are insensitive to the benzimidazoles, as are higher plants. The reason for the distinct differences in sensitivity is unknown but probably depends on single structural differences of the microtubule binding site. Resistance of this type, if stable, spreads rapidly and results in catastrophic disease control failures. The occurrence of resistance to the benzimidazoles is a controlling factor in their practical use and commercial value. However, not all targets are affected and because of their spectrum of activity, the benzimidazoles remain an important commercial fungicide group.

The widespread use of the benzimidazole fungicides is due to their high systemic activity against a wide range of Ascomycetes, Deuteromycetes and Basidiomycetes. The group is well-explored. Fuberidazole was first prepared in 1936 but not exploited as a fungicide until 1968. The main development began in the 1960s, and led to the identification of several important compounds, including benomyl, carbendazim, thiophanate-methyl and thiabendazole (Figure 4.14).

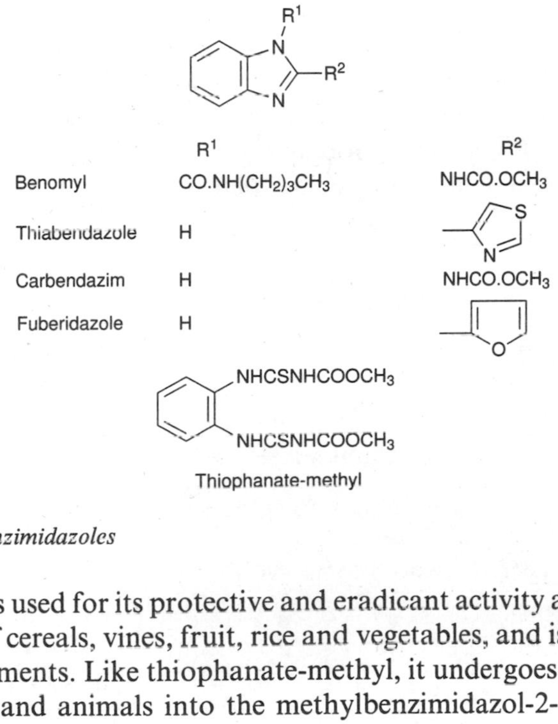

	R^1	R^2
Benomyl	$CO.NH(CH_2)_3CH_3$	$NHCO.OCH_3$
Thiabendazole	H	
Carbendazim	H	$NHCO.OCH_3$
Fuberidazole	H	

Thiophanate-methyl

Figure 4.14 *Benzimidazoles*

Benomyl is used for its protective and eradicant activity against several pathogens of cereals, vines, fruit, rice and vegetables, and is used in post-harvest treatments. Like thiophanate-methyl, it undergoes conversion in plants, soils and animals into the methylbenzimidazol-2-yl carbamate, carbendazim, which has a similar spectrum of activity.

Phenylcarbamates, for example, diethofencarb (Figure 4.15) are active against fungi that are resistant to benzimidazoles. The two groups may

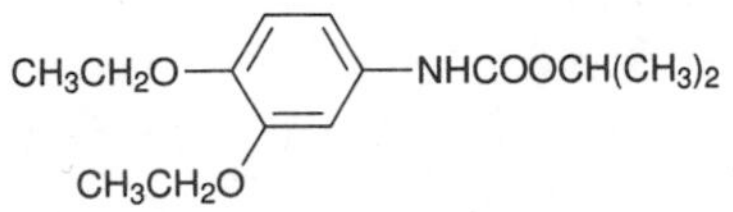

Figure 4.15 *Diethofencarb*

bind in the same region on the β-tubulin protein but their exact relationship is not fully defined.

4.6 Chitin Biosynthesis Inhibition

Chitin is the microfibrillar component of some fungal cell walls and is equivalent to cellulose microfibrils in plants. It is a characteristic feature of the Ascomycetes, Deuteromycetes and Basidiomycetes but is absent in the Phycomycetes, which contain cellulose as their major cell wall constituent.

Chitin molecules are $\beta(1,4)$-linked polymers of *N*-acetyl-D-glucosamine (Figure 4.16). The family of polyoxin fungicides obtained as secondary metabolites from *Streptomyces cacaoi* var. *asoensis* are structurally similar to UDP-*N*-acetyl-D-glucosamine and are competitive inhibitors of chitin synthase, which mediates in the process of transfer of *N*-acetyl-D-glucosamine from UDP-*N*-acetyl-D-glucosamine to the chitin polymer (Figure 4.17; Gooday[11]).

Chitin synthase is localised in the plasma membrane at the apex of

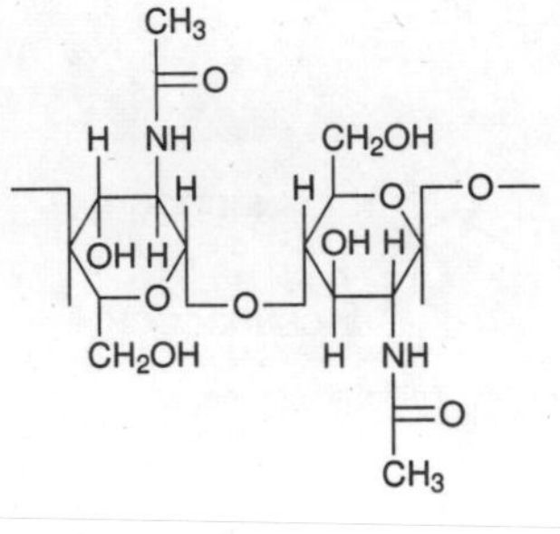

Figure 4.16 *Chitin β(1,4)-linked repeating unit*

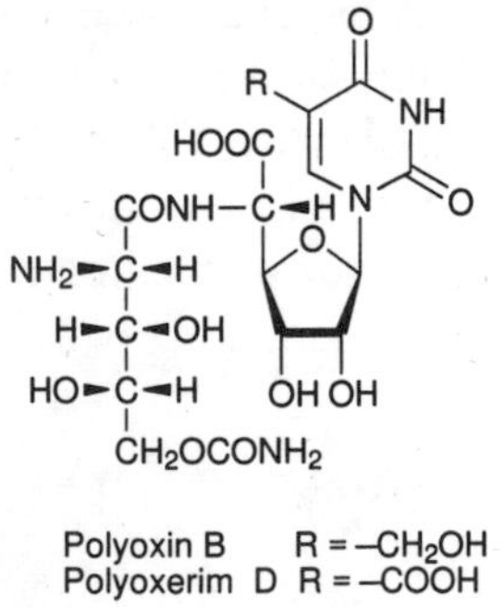

Polyoxin B   R = –CH₂OH
Polyoxerim D R = –COOH

Figure 4.17 *Polyoxins*

developing hyphae and its inhibition induces the collapse of cell wall integrity, visible as swelling and bursting of hyphal tips and spore germ tubes, and results in the eventual death of the fungus.

Polyoxins are used in the control of several pathogens, notably sheath blight of rice (*Rhizoctonia solani*). Polyoxin B is specific to *Alternaria*, *Botrytis*, and powdery mildew. Differences in activity between members of the polyoxin group probably reflect dissimilar uptake characteristics, which are governed by specific peptides. Different peptides may limit polyoxin activity at the level of the uptake mechanism and may also determine their resistance traits.

4.7 Melanin Biosynthesis Inhibition

Melanin is a collective term for brown or black pigments. They are condensates of phenolics and are widely distributed in nature, examples of this diverse chemistry being found in animals, plants and fungi. They perform various roles but in some fungi, notably *Pyricularia oryzae* and *Colletotrichum* spp., melanin biosynthesis is an essential feature in pathogenicity, the biosynthesis of melanin in appressorial walls being a requirement for the development of infection hyphae and subsequent penetration of the host epidermis. For example, melanin-deficient mutants of *P. oryzae* are not pathogenic.

Three compounds have been developed that utilise the inhibition of melanin biosynthesis as the basis of disease control (Figure 4.18). The first and most valuable of these is tricyclazole. Tricyclazole is readily absorbed by the foliage and roots of rice plants and redistributed acropetally. On treated plants, conidial germination of *P. oryzae* and

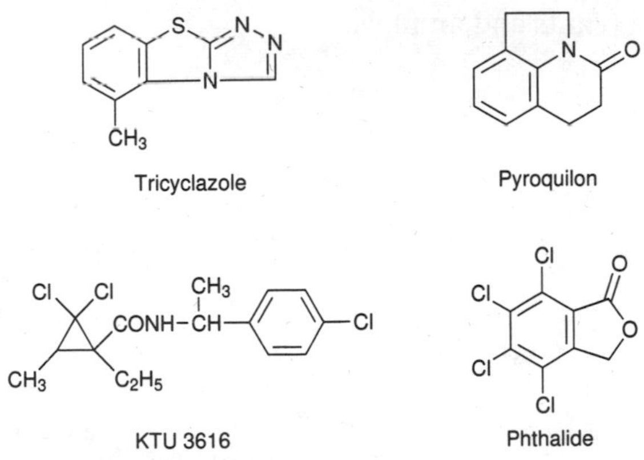

Tricyclazole

Pyroquilon

KTU 3616

Phthalide

Figure 4.18 *Melanin biosynthesis inhibitors*

the development of appressorial structures are unaffected but melanisation of the appressoria is reduced, changing in appearance from dark grey to light brown. The subsequent formation of the infection peg within the appressorium is inhibited, thereby blocking, the infection of the host.

Tricyclazole is reported to inhibit the melanin biosynthesis pathway between 1,3,8 trihydoxynaphthalene and vermelone whereas chromatographic analysis of *P. oryzae* culture media indicates that KTU 3616 treatment stimulates the accumulation of the intermediate, scytalone.

4.8 Protein Biosynthesis Inhibition

Protein biosynthesis involves the translation of the information contained in a base sequence in a specific mRNA into the synthesis of an amino acid sequence, or peptide.

There are two commercial fungicides, the antibiotics blasticidin S and kasugamycin, that act via the inhibition of protein biosynthesis (Figure 4.19). Blasticidin S is a fermentation product obtained from cultures of *Streptomyces griseochromogenes,* and has specific activity in the control of *P. oryzae*, similar to kasugamycin, a secondary metabolite of *S. kasugaensis*. However, much of the earlier work on mode of action was carried out using another antibiotic, cycloheximide.

Cyclohexamide prevents the incorporation of amino acids into protein through its affinity with the large subunit of the ribosome. Blasticidin S also interacts with the larger ribosomal subunit and blocks the binding site for aminoacyl-t-RNA, preventing the elongation of the protein chain. In contrast, kasugamycin binds to the smaller ribosomal subunit and inhibits protein elongation at lower concentrations than blasticidin S. Both antibiotics possess moderate systemic activity. They may have activity against plants and animals.

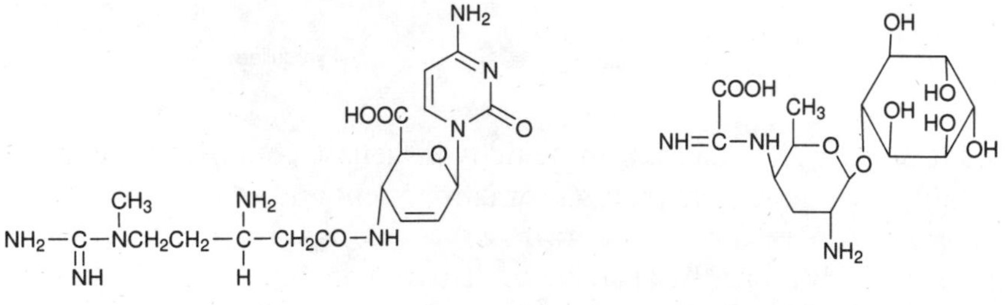

Figure 4.19 *Protein synthesis inhibitors*

4.9 Respiration Inhibition

Respiration is a ubiquitous and complex oxidative mechanism that conserves energy, liberated by the catabolism of molecules such as carbohydrates, through the production of ATP from ADP and orthophosphate. Several component reaction sequences can be identified.

Acetyl-CoA formed via the breakdown of carbohydrates (in glycolysis), proteins and fats is degraded by the tricarboxylic acid (TCA) cycle into carbon dioxide. In the process, the co-factor NAD^+ (nicotinamide adenine dinucleotide) and FAD (flavin adenine dinucleotide), the flavoprotein component of the succinate dehydrogenase enzyme complex in the TCA cycle, are reduced to form NADH and $FADH_2$, respectively. The oxidised co-factors are regenerated by surrendering their electrons to a cascade of respiratory carriers terminating in the reduction of oxygen to water. The electron transport chain is coupled to the production of ATP through the process of oxidative phosphorylation, and complements a smaller amount of ATP synthesised during the earlier reactions of glycolysis.

Respiratory inhibitors not only deprive the organism of the major end-product, ATP, but may interfere with the production of other vital synthetic intermediates. Several sites of action have been identified for fungicides.

The electron transport chain can be fractionated into four components, each associated with a characteristic activity (Figure 4.20).

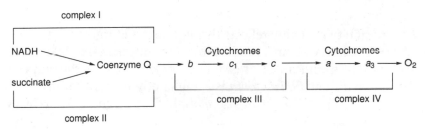

Figure 4.20 *Mitochondrial electron transport*

4.9.1 Complex II. The succinate dehydrogenase complex, or 'complex II', contains a non-haem iron-sulfur component and utilises the electron acceptor FAD to effect the transfer of electrons from $FADH_2$ to coenzyme Q. Inhibitors of succinate dehydrogenase are specific basidiomycete fungicides, with uses against smuts, bunts, and *Rhizoctonia* spp. (Figure 4.21).

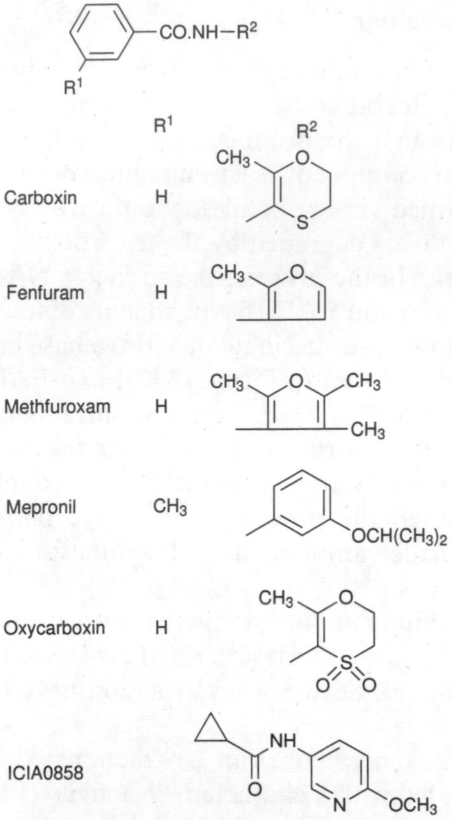

Figure 4.21 *Carboxamides*

The specificity probably reflects the unique structural configuration of the target site in Basidiomycetes.[12] However, differential uptake between species may be significant. For example, the uptake of carboxamides by *Ustilago maydis* and *Rhizoctonia* is significantly higher than by insensitive species such as *Fusarium oxysporum* f. sp. *lycopersici*. However, differences in uptake between species are not observed consistently. In the case of ICIA0858, a carboxamide from Zeneca Agrochemicals, variety differences between wheat crops correlate with fungicidal activity, suggesting a role for host metabolism in selectivity.

There are many compounds in the carboxamide group. Carboxin, the lead product, is used extensively as a seed treatment in a wide variety of crops. Fenfuram and methfuroxam, produced by the substitution of the 1,4-oxathiin ring of carbon by a furan moiety, are used for the control of seed-borne pathogens in cereals. Similarly, mepronil is used to control sheath blight in rice and rust diseases in wheat.

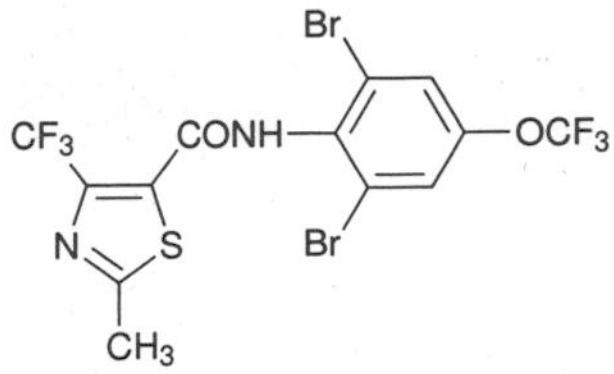

Figure 4.22 *Thifluzamide (MON 24000)*

Thifluzamide (MON 24000) is the most recent addition to the carboxamide group. It is a thiazolecarboxanilide, with activity against a wide spectrum of foliar and seed-borne pathogens (Figure 4.22). Thifluzamide also demonstrates some interesting phloem-systemic activity against take-all of cereals, *Gaeumannomyces graminis* f. sp. *tritici*. If proven, this will be a major advance in the control of this widespread and damaging soil pathogen.

4.9.2 Complex III. The development of this group of fungicides is new and will compete directly with the azoles, which dominate the largest fungicide markets. In *in vitro* enzyme assays, inhibitory concentrations for this group are low, the most recent, famoxadone, being up to two orders of magnitude more active than earlier representatives (Table 4.1).

Table 4.1 *Comparative* in vitro *activity of mitochondrial electron transport (NADH to O_2) by complex III inhibitors.*

Compound	$IC_{50}(ppb)$
Famoxadone	4.5
Azoxystrobin	170.0
Kresoxim-methyl	75.0

The development of the strobilurins as agricultural fungicides began in the 1960s following the discovery of a natural fungicide in the basidiomycete *Strobilurus tenacellus*. Another basidiomycete fungus, *Oudemansiella mucida*, was shown to produce a similar fungicidal secondary metabolite.

By the early 1980s, the *in vitro* activity of strobilurin A, the compound from *Strobilurus tenacellus,* and oudemansin A, from *Oudemansiella mucida*, had been completely characterised[13] (Figure 4.23) and the mode of action was known to be the inhibition of electron transfer in complex

Strobilurin A   Oudemansin A

Figure 4.23

III of mitochondrial respiration.[14] The compounds exhibited good broad-spectrum activity *in vitro*, comparable with synthetic standards, but *in vivo* was inadequate to justify further development as a fungicidal product. Further study showed that activity was lost due to photolytic instability, and led to the production (Figure 4.24) of a series of analogues and selection of the eventual product candidates kresoxim-methyl (BAS 490 F)[15] and azoxystrobin (ICIA5504).[16]

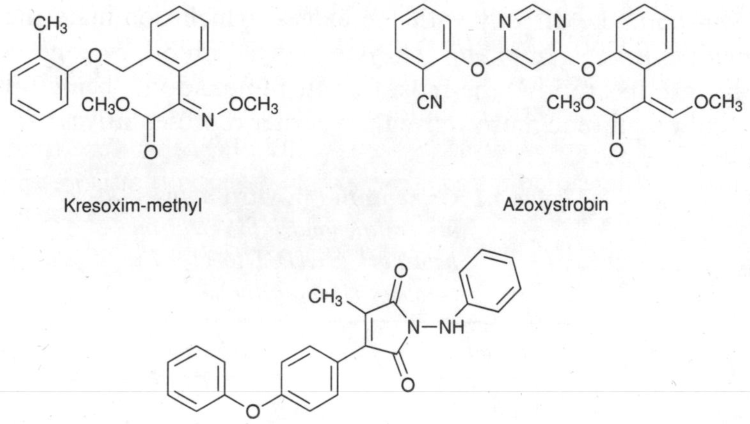

Kresoxim-methyl   Azoxystrobin

Famoxadone

Figure 4.24 *Inhibitors of complex III*

The spectrum is unusually broad, including representatives of all fungal classes in major crop targets. Generally, the compounds are systemic and provide long-term disease control though their redistribution within the crop via a continuous mechanism of absorption into the xylem from the waxy cuticular layer of leaves. Characteristically, spore germination occurring on the leaf surface is the most sensitive developmental stage of target fungi.

4.9.3 Oxidative Phosphorylation.

Oxidative phosphorylation, that is the production of ATP during the passage of electrons down the terminal electron transport chain, may be disrupted in two distinct ways. Compounds that divorce the process of electron transport and the phosphorylation of ADP are termed uncoupling agents. They permit NADH and succinate to be oxidised via the electron transport chain without the production of ATP and are lethal. Oxidative phosphorylation may also be inhibited directly, thus preventing the oxidation of NADH and succinate. Several products are available that exploit these modes of action. Characteristically, they have wide activity spectra that span major disciplines of pesticide use.

Nitrophenols are uncoupling agents. The structure of dinitrophenols conforms to the generalised model for uncoupling agents. They are lipid-soluble and possess an acidic moiety and an aromatic ring. To operate, oxidative phosphorylation relies upon the maintenance of an H^+ gradient across the inner mitochondrial membrane. Nitrophenols disrupt the lipid matrix of mitochondrial membranes and allow H^+ ions to equilibrate, thus removing the gradient.

2,4-Dinitrophenol was the first uncoupling agent to be discovered. The first organic fungicide to compete with traditional sulfur-based materials in the control of powdery mildews was the nitrophenol, dinocap. Dinocap undergoes hydrolysis following uptake to produce toxic dinitrophenols. Other uncoupling agents include nitrothal-isopropyl and fluazinam (Figure 4.25).

The triphenyltins fentin acetate and fentin hydroxide are non-systemic

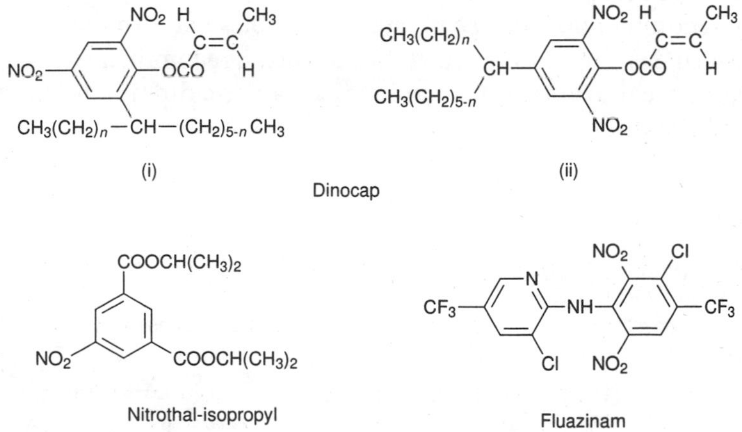

(i)   (ii)

Dinocap

Nitrothal-isopropyl

Fluazinam

Figure 4.25 *Nitrophenols*

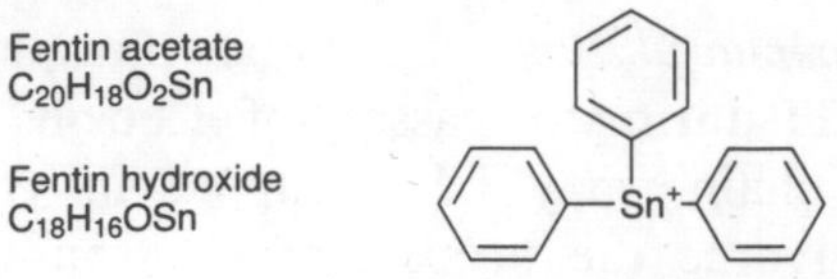

Fentin acetate
$C_{20}H_{18}O_2Sn$

Fentin hydroxide
$C_{18}H_{16}OSn$

Figure 4.26 *Fentins*

fungicides (Figure 4.26). They are used in potatoes, rice and vegetables for the control of several pathogens, but are potentially phytotoxic.

Fentin hydroxide has been demonstrated to inhibit oxidative phosphorylation in rat liver mitochondria and it is thought that the membrane-located component of ATPase is the target for triphenyltin fungicides.

4.10 Interference with Cell Membrane Structure

Cell membranes are bilayers of amphipathic acids, for example phospholipids and sterols, which contain globular proteins. The structure is governed by the essential requirement for stability in an aqueous environment, that is, the hydrophobic tails of the lipid molecules point towards each other, leaving the outer surfaces composed of polar, hydrophilic groups.

The guanidines, comprising dodine and guazatine, have long-chain alkyl groups and act as non-specific detergents. The lipophilic alkyl chain attaches to the lipid fraction of membranes whilst the polar guanidino portion remains in the aqueous phase. The result is a disruption in the membrane permeability characteristics and active transport systems.

Dodine (Figure 4.27) is a protectant fungicide, with some eradicant action, recommended for use in fruit, vegetables and ornamentals.

Guazatine is a mixture of guazatine acetates, recommended as a broad spectrum cereal seed treatment, a foliar spray or a dip treatment for fruit and seed potatoes.

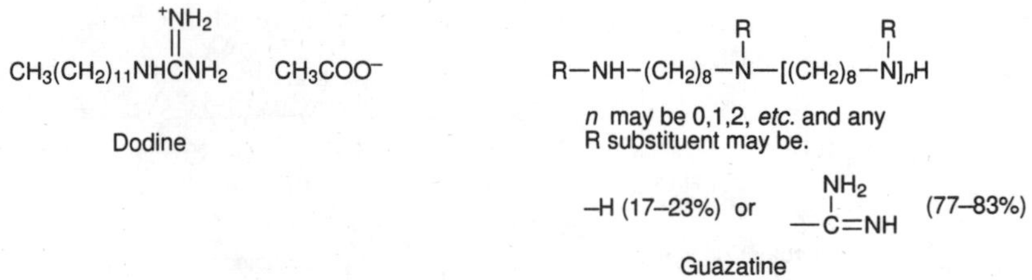

$$CH_3(CH_2)_{11}NH\overset{\overset{+NH_2}{\|}}{C}NH_2 \quad CH_3COO^-$$

Dodine

$$R-NH-(CH_2)_8-\overset{\overset{R}{|}}{N}-[(CH_2)_8-\overset{\overset{R}{|}}{N}]_nH$$

n may be 0,1,2, *etc.* and any R substituent may be.

$-H$ (17–23%) or $\quad -\overset{\overset{NH_2}{|}}{C}=NH \quad$ (77–83%)

Guazatine

Figure 4.27 *Guanidines*

4.11 Undefined Mode of Action

Many commercial fungicides have unknown or indistinct modes of action. This in no way detracts from their commercial potential but does delay the development of particular chemistries through directed synthesis.

4.11.1 Anilinopyrimidines. This group, also known as pyrimidin-amines, is recently developed and comprises three compounds (Figure 4.28), mepanipyrim, pyrimethanil and cyprodinil. They have potential use in a wide variety of crops and are broad-spectrum fungicides. Mepanipyrim and pyrimethanil are effective against fruit pathogens, especially *Botrytis cinerea* on vine and *Venturia* spp. on top fruit. Cyprodinil is also active against several cereal pathogens, notably *Pseudocercosporella herpotrichoides* (eyespot).

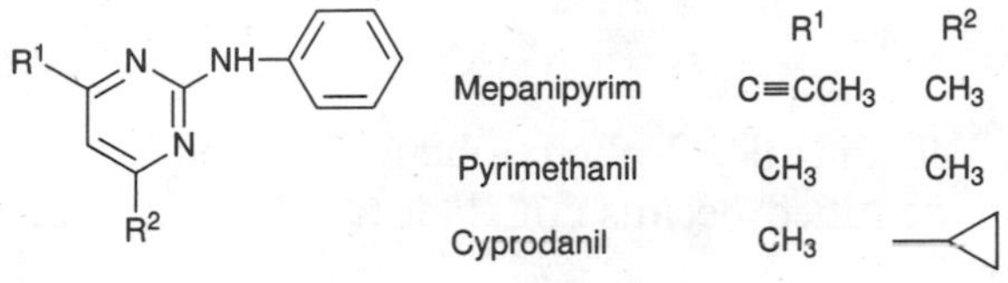

	R^1	R^2
Mepanipyrim	C≡CCH$_3$	CH$_3$
Pyrimethanil	CH$_3$	CH$_3$
Cyprodanil	CH$_3$	◁

Figure 4.28 *Anilinopyrimidines*

The inhibitory effects of pyrimethanil on spore germination and germ tube extension are limited and appressorium production and host penetration processes are unaffected. However, a protectant application of pyrimethanil results in a significant reduction in the number of host cells killed at each penetration site. The major effect of pyrimethanil may be through a reduction in the secretion of fungal enzymes either by a reduction in the biosynthesis of lytic enzymes or their secretion into the penetration court. Enzymes involved with cell lysis include pectinases, cellulases, proteinases and laccase.

However, pyrimethanil and mepanipyrim do not inhibit proteinase, cellulase or polygalacturnase activity in *Botrytis cinerea*[17] but reduce pectinase and invertase secretion with an associated increase in their intracellular accumulation. This is proposed to be the mechanism of action of the anilinopyrimidines but the biochemical basis of the effect is not known. There is evidence that suggests the involvement of methionine biosynthesis inhibition.[18]

4.11.2 Aromatic Hydrocarbons and Related Structures. This group of fungicides includes hexachlorobenzene, quintozene, tecnazene, chloro-

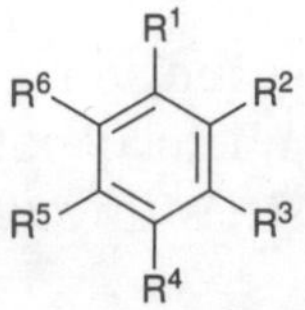

	R^1	R^2	R^3	R^4	R^5	R^6
Hexachlorobenzene	Cl	Cl	Cl	Cl	Cl	Cl
Quintozene	NO$_2$	Cl	Cl	Cl	Cl	Cl
Tecnazene	NO$_2$	Cl	Cl	H	Cl	Cl
Chloroneb	OCH$_3$	Cl	H	OCH$_3$	Cl	H
Dicloran	NO$_2$	Cl	Cl	Cl	NH$_2$	Cl

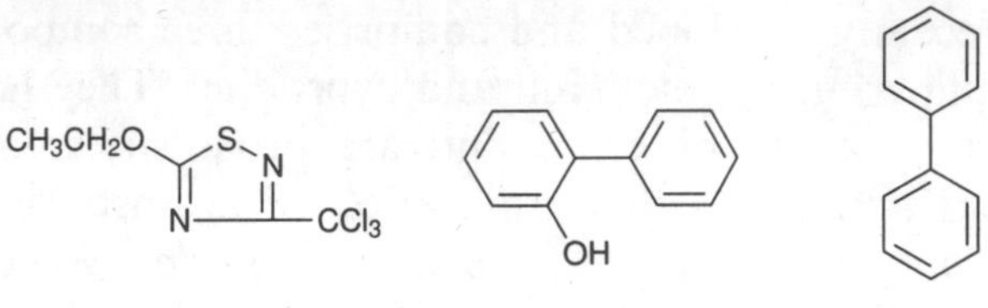

Etridiazole   2-Phenylphenol   Biphenyl

Figure 4.29 *Aromatic hydrocarbon fungicides*

neb, dicloran, etridiazole, 2-phenylphenol and biphenyl (Figure 4.29) and were first identified because of their ability to induce mutations in fungal colonies.

These fungicides are predominantly applied to control soil and seed-borne diseases in cotton, vegetables, sugar beet and ornamentals, as well as post-harvest fungi. Although some systemic activity has been reported, redistribution is largely through the vapour phase.

The compounds act fungistatically rather than fungicidally and inhibit mycelial growth but not spore germination. Symptoms include the swelling of hyphal tips and germ tubes, often resulting in lysis. Microscopy indicates that lipid peroxidation is the primary toxic target in sensitive fungi. Mitochondria exhibit characteristic swelling of the cristae and disruption of the inner mitochondrial membrane and outer membrane. Vacuolisation of the nuclear envelopes is accompanied by the breakdown of the endoplasmic reticulum and plasmalemma. Cell wall thickening often occurs and may be related to the removal of chitin and glucan inhibitor proteins in the plasmalemma, leading to a stimulated production of cell wall components.

4.11.3 Dicarboximides. Dichlozoline was the first reported dicarboximide and demonstrated potentially useful activity against *Sclerotinia* and *Botrytis*. This compound is no longer available but several related compounds have been launched, including iprodione, vinclozolin, procymidone and chlozolinate (Figure 4.30).

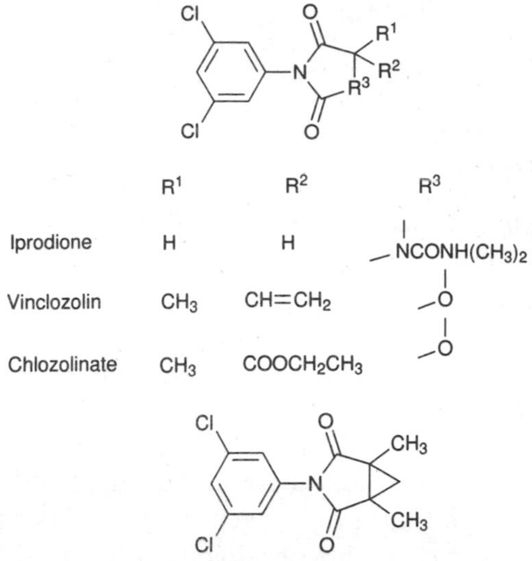

Figure 4.30 *Dicarboximides*

Their development and commercial success was favoured by the coincidence of resistance to benzimidazole fungicides.

The range of activity of the dicarboximides includes the control of many major ascomycete, deuteromycete and basidiomycete pathogens in a wide range of crops. The activity spectrum varies slightly between compounds.

The dicarboximides inhibit spore germination and cause increased branching, swelling and lysis of germ tubes and hyphal tips. Effects on cell division have been reported but no major inhibition of nucleic acid metabolism, respiration, protein or lipid synthesis has been observed.

Dicarboximide and aromatic hydrocarbon fungicides exhibit cross resistance, suggesting that their primary mode of action is via their effects on lipid peroxidation. Treatment with dicarboximides at concentrations that also affect mycelial development specifically inhibit cytochrome *c* reductase, blocking electron transfer from NADPH to cytochrome *c*. However, recent studies have highlighted the similarity between the phenylpyrrole fungicides, fenpiclonil and fludioxonil, and the dicarbox-imides, with respect to their cross-resistance patterns in laboratory-generated mutants of *Fusarium sulfureum*, suggesting that the mode of action may be based on the inhibition of transport-associated phosphor-ylation of glucose.[19] The situation is unclear because field isolates of *Botrytis cinerea* that are resistant to dicarboximides are highly sensitive to the phenylpyrroles.

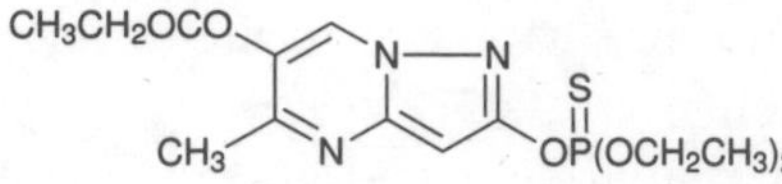

CH₃CH₂OCO N—N
          S
CH₃ N   OP(OCH₂CH₃)₂

Figure 4.31 *Pyrazophos*

4.11.4 Organophosphorus Fungicides. This group comprises pyrazophos (Figure 4.31) and the phosphonic acid amides, triamiphos and ditalimfos (superseded compounds).

Pyrazophos has its main uses against powdery mildews in which it inhibits conidial germination and appressorium formation. The active fungicidal agent is 2-hydroxy-5-methyl-6-ethoxycarbonylpyrazolo-(1,5-α)-pyrimidine formed through the action of microsomal mixed function oxidases. Several modes of action are suggested, including interference in protein synthesis, inhibition of oxygen uptake, disruption of fatty acid metabolism and reduced phospholipid synthesis affecting membrane structures and function, and the inhibition of melanin synthesis.

4.11.5 Fosetyl. Fosetyl (aluminium ethyl phosphate; Figure 4.32) is a specific downy mildewicide (Oomycetes) used to control *Plasmopara viticola* in vines and *Phytophthora* blights in various fruit and nut crops.

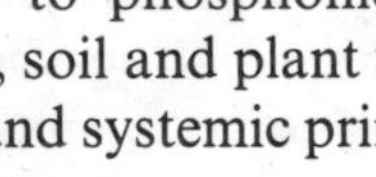

$$\left(CH_3CH_2O-\underset{H}{\overset{\overset{\displaystyle O}{\|}}{P}}-O \right)_3 Al$$

Figure 4.32 *Fosetyl*

Fosetyl readily degrades to phosphonic acid (H_3PO_3) and carbon dioxide in aqueous solution, soil and plant tissue and phosphonic acid is considered to be the active and systemic principle through the disruption of phospholipid metabolism.

4.11.6 Prothiocarb, Propamocarb. Prothiocarb (superseded compound) and its oxygen analogue propamocarb (Figure 4.33) are selectively active against Oomycetes. Propamocarb, as its hydrochloride, is xylem mobile with uses as a seed treatment, drench, soil incorporation, dip or foliar spray in the control of diseases caused by Oomycetes on ornamentals, tobacco, fruit and potatoes.

$(CH_3)_2NCH_2CH_2CH_2NHCOOCH_2CH_2CH_3.HCl$

Figure 4.33 *Propamocarb*

The action of propamocarb is related to membrane function, causing efflux of cell constituents. Leakage ceases after the development of mycelium and can be inhibited by the addition of sterols.

4.11.7 Cymoxanil. Cymoxanil (Figure 4.34) is a systemic urea with protectant and curative activity against Oomycetes, particularly *Plasmopara viticola* on vine and *Phytophthora infestans* (potato late blight).

$$CH_3CH_2NHCONHCOC\overset{\displaystyle CN}{=}NOCH_3$$

Figure 4.34 *Cymoxanil*

Mycelial growth is more sensitive to cymoxanil than early growth phases, including the release of zoospores from sporangia and their germination. Cymoxanil inhibits nucleic acid and protein synthesis in some fungi but may have to be activated to induce a fungicidal response.

4.11.8 Quinoxyfen (DE-795). Quinoxyfen (DE-795), announced by Dow AgroScience in 1996 (Figure 4.35), is a specific powdery mildew fungicide with particular utility in cereals. The mode of action is unknown but the chemistry is close to LY214352, a known inhibitor of pyrimidine synthesis. Quinoxyfen and RNA-synthesis fungicides also inhibit appressorium formation.

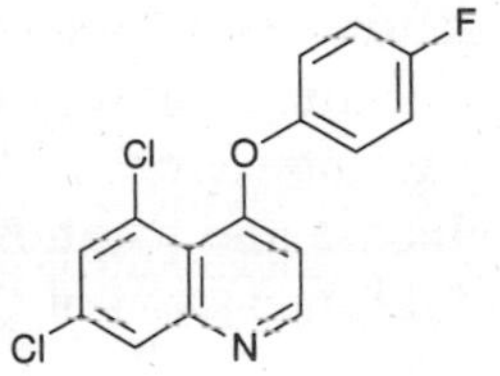

Figure 4.35 *Quinoxyfen (DE-795)*

4.11.9 Dimethomorph. Dimethomorph (Figure 4.36) is effective against Oomycetes except *Pythium* spp. The compound, which is a cinnamic acid derivative, operates as a protectant but also has some curative activity that can be modified by means of formulation. Only the Z isomer is active but because of the rapid interconversion of isomers in the light, there is no practical advantage in its specific synthesis.

Dimethomorph inhibits the formation of the fungal cell wall and is not cross-resistant to any known class of fungicides.

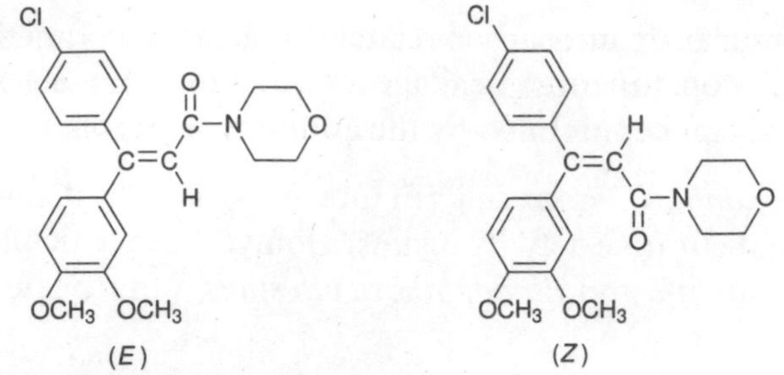

Figure 4.36 *Dimethomorph*

4.11.10 Phenylpyrroles. The stabilisation of the light-labile compound pyrrolnitrin, a secondary metabolite from *Pseudomonas pyrrocina*, led to the discovery of fenpiclonil and fludioxonil (Figure 4.37). The range of fungicidal activity is broad, including all fungal classes except the Phycomycetes.

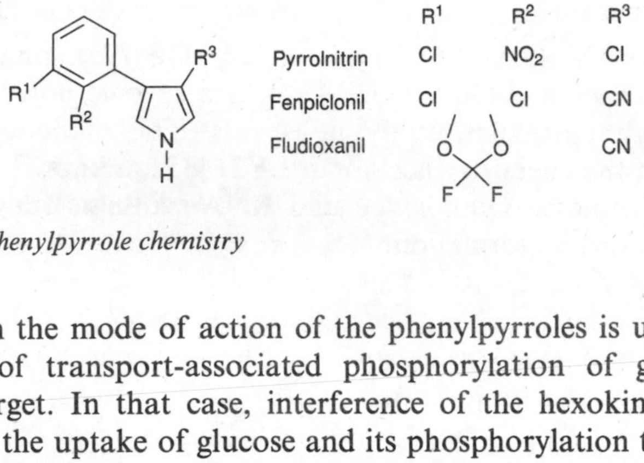

	R^1	R^2	R^3
Pyrrolnitrin	Cl	NO_2	Cl
Fenpiclonil	Cl	Cl	CN
Fludioxanil			CN

Figure 4.37 *Phenylpyrrole chemistry*

Although the mode of action of the phenylpyrroles is uncertain the inhibition of transport-associated phosphorylation of glucose is a possible target. In that case, interference of the hexokinase-complex controlling the uptake of glucose and its phosphorylation to glucose-6-phosphate leads to the gradual starvation of the fungus and its eventual death.

4.12 Interference with Disease Recognition Mechanisms

There is an argument for considering the disease condition as the expression of symptoms rather than as the state of merely hosting an alien organism. In addition, the degree and timing of symptom development can be regarded as a measure of compatibility of the host and the alien fungus. At one extreme, both are able to survive in balance for the whole of their lives and in some cases, the associations are so beneficial that selection favours mutual dependence, as in mychorrhizal partnerships and lichens. However, since organisms at this level of complexity are not altruistic, such associations can be considered to be cases of equally balanced parasitism. A slight imbalance towards the benefit of the fungal partner results in long-term, often obligate, diseases in which precise host specificity by the fungus is commonplace and the expression of symptoms is delayed far beyond the moment of host colonisation. Examples are the smut diseases of temperate cereals in which the invasion of the seed and the developing plant by the fungus remains unrecognised by the host, and hence asymptomatic, until the production of seeds. Less prolonged examples are the rust and *Septoria* diseases of cereals. At the other extreme of this continuum of recognition are the necrotrophic fungi that only invade the plant through the colonisation of dead tissue, killed in advance of the expanding fungal colony by the action of excreted enzymes. How hosts recognise fungal challenges is uncertain but the mechanisms that follow recognition are well-known. They include responses such as:

- the formation of physical barriers and the production of antifungal chemicals;
- hypersensitivity (localised death of plant cells);
- systemic acquired resistance (SAR).

The effectiveness of these defence mechanisms is clear from the ensuing development of the invading fungus and from the fact that most plants are immune to most diseases. Commercial success in manipulating these responses is confined to the induction or mimicry of SAR. Other approaches have generally shown less promise. For example, the induction of chemicals (phytoalexins) that promote hypersensitivity manifests itself as a herbicidal rather than a fungicidal response since such chemicals limit fungal attack by destroying adjacent host tissue.

4.12.1 Systemic Acquired Resistance. SAR is the induction of a transient long-distance (translocated) defence response to fungal attack that is distinct from the local production of phytoalexins. The transient nature of SAR is important and has likely been selected in the course of evolution as a corollary to the energy demands that are made on the plant by the permanent mobilisation of resistance mechanisms. However, plants challenged by fungi are sensitised to subsequent attack and the speed of their SAR response is increased.

The phenomenon is well-known[20] and correlates with the accumulation of β-1,3 glucanases and chitinases, the cell wall degrading enzymes excreted by some fungi. Chemicals that promote SAR are potential control agents with long-term persistence. However, such chemical inducers have a limited spectrum, the range of sensitive pathogens being equivalent to those controlled by the naturally induced response.

Salicylic acid is an inducer of antifungal proteins (Figure 4.38) and is proposed to activate the expression of SAR genes in the plant.[21]

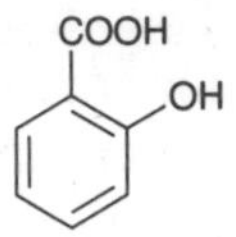

Figure 4.38 *Salicylic acid*

CGA 245704 (acibenzolar; Figure 4.39) may operate in a similar manner to natural defence activators. It does not possess fungicidal activity *per se* but induces resistance to a wide range of pathogens. The product is readily translocated in the phloem and xylem but is subject to rapid metabolism.

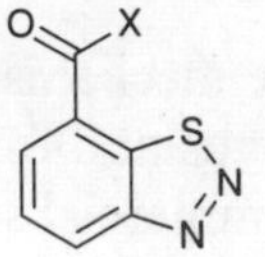

X = (1) SCH₃ (CGA 245704); (2) OCH₃ (initial lead)

Figure 4.39 *Benzothiadiazoles*

Probenazole is used for the control of *Pyricularia oryzae* in rice (Figure 4.40). Interestingly, probenazole is ineffective against *P. oryzae* as a pathogen of barley, implying a dependence upon the metabolism of the natural host, rice, or a different SAR response.

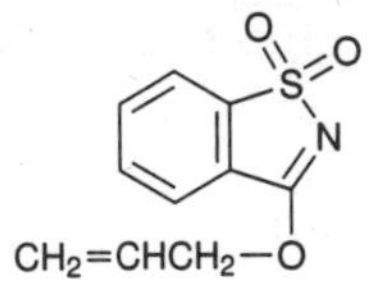

CH₂=CHCH₂—O

Figure 4.40 *Probenazole*

5 FUNGICIDE RESISTANCE ACTION COMMITTEE (FRAC)

FRAC is an industry-based organisation comprising a Central Steering Committee and several Working Groups whose collective objective is to anticipate and manage potential and actual resistance problems arising from the use of fungicides. Currently there are six Working Groups (FRAC 1996 Status Report) that establish and direct collaborations on:

- monitoring programmes to establish baseline data, interpret results and standardise methodologies;
- the definition of research objectives to enhance the decision-making processes of FRAC;
- educational programmes;
- verification of information on resistance, including investigations, remedies and statements from industry and academia;
- recommendation of anti-resistance strategies;
- liaison with the activities of national and regional sub-groups.

The groups are focused on major areas of fungicide chemistry that have presented themselves as real or potential resistance risks in agriculturally important crops. These are:

- sterol biosynthesis – azoles/morpholines/piperidines;
- phenylamides – metalaxyl/oxadixyl/ofurace/benalaxyl;
- dicarboxamides – vinclozolin/procymidone/iprodione;
- benzimidazoles – benomyl/carbendazim/thiabendazole/thiophanate/thiophanate-methyl;
- anilinopyrimidines – cyprodanil/pyrimethanil/mepanipyrim;
- strobilurins – azoxystrobin/kresoxim-methyl/famoxadone.

6 QUESTIONS

4.1 Can a compound like mancozeb be used as a systemic fungicide?
4.2 Why are systemics more popular than non-systemics?
4.3 What is the basis for the selectivity of polyoxins?
4.4 What is the major cause of resistance to fungicides?
4.5 Why do SAR fungicides have a limited spectrum?
4.6 Growers have a definite view on what an ideal fungicide should be. What would be yours and why?

7 REFERENCES

1. E.C. Large, 'The Advance of the Fungi', Jonathan Cape, London, 1958.

2. E.L. Mercer, 'Sterol Biosynthesis Inhibitors: their current status and modes of action', *Lipids*, 1991, **26**, 584–597.

3. B.C. Baldwin and A.J. Corran, 'Inhibition of Sterol Biosynthesis: application to the agrochemical industry', in 'Antifungal Agents. Discovery and Mode of Action', eds. G.K. Dixon, L.G. Copping and D.W. Hollomon, BIOS Scientific Publishers Ltd., Oxford, 1995, pp. 59–68.

4. R.I. Baloch, E.I. Mercer, T.E. Wiggins and B.C. Baldwin, 'Inhibition of Ergosterol Biosynthesis in *Saccharomyces cerevisiae* and *Ustilago maydis* by Tridemorph, Fenpropimorph and Fenpropidin', *Phytochemistry*, 1984, **23**, 2219–2226.

5. W. Köller, 'Antifungal Agents with Target Sites in Sterol Function and Biosynthesis', in 'Target Sites of Fungicide Action', ed. W. Köller, CRC Press, Boca Raton, FL, 1992, pp. 119–206.

6. G.D. Robson and A.J.P. Trinci, 'Inhibition of Phospholipid and Phosphoinositide Biosynthesis', in 'Antifungal Agents. Discovery and Mode of Action', eds. G.K. Dixon, L.G. Copping and D.W. Hollomon, BIOS Scientific Publishers Ltd, Oxford, 1995, pp. 77–94.

7. I. Yamaguchi, 'Pesticides of Microbial Origin and Applications of Molecular Biology', in 'Crop Protection Agents from Nature: natural products and analogues', ed. L.G. Copping, Royal Society of Chemistry, Cambridge, UK, 1996, pp. 20–49.

8. L.G. Davidse and G.C.M. van der Berg-Valthius, 'Biochemical and Molecular Aspects of the Phenylamide Fungicide–Receptor Interaction in Plant Pathogenic *Phytophthora* spp.', in 'Signal Molecules in Plants and Plant–Microbe Interactions', ed. B.J.J. Lugtenberg, Springer-Verlag, Berlin, 1989, pp. 261–278.

9. G. Gustafson, 'Nucleic Acid Metabolism as a Target for Antifungals: the mechanism of action of LY214352', in 'Antifungal Agents. Discovery and Mode of Action', eds. G.K. Dixon, L.G. Copping and D.W. Hollomon, BIOS Scientific Publishers Ltd., Oxford, 1995, pp. 111–117.

10. L.G. Davidse and W. Flach, 'Differential Binding of Benzimidazol-2-yl Carbamate to Fungal Tubulin as a Mechanism of Resistance to this Antimitotic Agent in Mutant Strains of *Aspergillus nidulans*', *J. Cell. Biol.*, 1977, **72**, 174–193.

11. G.W. Gooday, 'Cell Walls', in 'The Growing Fungus', ed. N.A.R. Gow and G.M. Gadd, Chapman & Hall, London, 1995, pp. 43–62.

12. H. Buchenauer, 'Physiological Reactions in the Inhibition of Plant Pathogenic Fungi', in 'Chemistry of Plant Protection. 6. Controlled Release, Biochemical Effects of Pesticides, Inhibition of Plant Pathogenic Fungi', eds. W.S. Bowers, W. Ebing, D. Martin and R. Wegler, Springer-Veslag, Berlin.

13. T. Anke, F. Oberwinkler, W. Steglich and G. Schramm, 'The Strobilurins – new antifungal antibiotics from the basidiomycete *Strobilurus tenacellus*, *J. Antibiot.*, 1997, **30**, 806–810.

14. W.F. Becker, G. Von Jagow, T. Anke and W. Steglich, 'Oudemansin,

Strobilurin A, Strobilurin B and Myxothiazole: new inhibitors of the bc_1 segment of the respiratory chain with an (E)-β-methoxyacrylate system as common structural elements', *FEBS Lett.*, 1981, **132**, 329–333.

15. E. Ammermann, G. Lorenz and K. Schelberger, 'BAS 490F – A broad spectrum fungicide with a new mode of action', 'Proceedings of the British Crop Protection Conference – Pests and Diseases', 1992, Vol. 1, 403–410.

16. J.R. Godwin, V.M. Anthony, J.M. Clough and C.R.A. Godfrey, 'ICIA5504, A Novel Spectrum, Systemic β-Methoxyacrylate Fungicide', 'Proceedings of the British Crop Protection Conference – Pests and Diseases', 1992, Vol. 1, pp. 435–442.

17. R.J. Milling, A. Daniels and J.B. Pillmoor, 'The Mode of Action of the Anilinopyrimidines: a new fungicide target', in 'Antifungal Agents. Discovery and Mode of Action', eds. G.K. Dixon, L.G. Copping and D.W. Hollomon, BIOS Scientific Publishers Ltd., Oxford, 1995, pp. 201–209.

18. P. Masner, P. Muster and J. Schmid, 'Possible Methionine Biosynthesis Inhibition by Pyrimidinamine Fungicides', *Pesticide Sci.*, 1994, **42**, 163–166.

19. P. Leroux, C. Lanae and R. Fritz, 'Similarities in the Antifungal Activities of Fenpiclonil, Iprodione and Tolclofos-methyl against *Botrytis cinerea* and *Fusarium nivale*', *Pesticide Sci.*, 1992, **36**, 255–261.

20. K.S. Chester, 'The Problem of Acquired Physiological Immunity in Plants', *Q. Rev. Biol.*, 1933, **8**, 275–324.

21. J.P. Metraux, H. Signer, J. Ryals, E.W. Ward, M. Wyss-Benz, J. Gaudin, K. Raschdorf, E. Schmid, W. Blum and B. Inverardi, 'Increase in Salicylic Acid at the Onset of Systemic Acquired Resistance in Cucumber', *Science*, 1990, **250**, 1004–1006.

CHAPTER 5

Plant Growth Regulators

1 INTRODUCTION

Crop plants are complex matrices of single cells organised into distinct structures such as roots, leaves and flowers. Each structure, or plant organ, is characterised by further complexities of differentiation at the cellular level. Thus, leaves consist of epidermal cells, including the specialised stomata, hydathodes and leaf hairs, palisade tissue, which is the centre for photosynthesis, and mesophyll cells. Some structures, for example, the long distance transport systems are found throughout the plant. These comprise cells that are developed as xylem and phloem elements and are responsible for the movement of water, minerals and organic molecules. Despite the considerable degree of cellular specialisation that accompanies the organisation of higher plants, most cells are capable of dedifferentiation under suitable conditions and can be used to produce other plants. This is the basis of plant cloning and the technique is used extensively in breeding and horticulture.

Dedifferentiation of plant tissue can sometimes be symptomatic of plant disease or pest infestations. Outgrowths of leaf and stem tissue appear as galls, common examples being oak apples and the phenomenon known as 'witches broom' that occurs on tree branches. Both are the result of damage, usually by insects. Although such growths represent a degree of deviation from the normal organised state of the plant, they can take on a characteristic form that is specific to the invading insect. In addition, many fungal pathogens can interfere with the normal development of plants, not only by causing the production of galls but also by exaggerating particular features. *Gibberella fujikuroi* is a pathogen of rice that causes a rapid elongation of stem tissue. In evolutionary terms, this is advantageous to the spread of the fungus since it raises it high above the crop and enables the effective dispersal of spores. Unfortunately for the rice plant, the result is that the plant loses stem strength and eventually collapses. Appropriately, the condition has acquired the name 'Bakanae disease', or 'foolish seedling disease'. On a smaller scale, rust diseases are well-known for their ability to produce 'green islands' around emergent uredospore pustules on infected leaves. These are formed by the inhibition of leaf tissue senescence by the fungus.

Clearly, cell growth, differentiation and organisation in plants and the maintenance of normal plant function is under a form of control and, importantly, it is susceptible to external influences. The control mechanism is chemical. Seeds germinate in response to suitable levels of temperature, water and light. The process of germination, however, is mediated by the hydrolysis of endospermic starch, mediated by a sterol, gibberellin. Similarly, subsequent geotropic growth patterns of the root and phototropic affinities of the shoot are controlled via the synthesis, transport and local accumulation of an indole. Several classic studies using excised plant parts, and others that investigated the responses of plants to damage and invasion by pathogens or pests, have defined a diverse family of substances implicated in growth regulation. They include the gibberellins, auxins, cytokinins, ethylene and abscisic acid. In addition to their major role in development, these so-called 'plant hormones' or 'plant growth substances' have a long-term function in adapting the plant to broad changes in the environment and, in the short-term, allow the plant to react to sudden or diurnal events. Similar to animal hormones, these are active in minute amounts, they generally act at a distance and are rapidly destroyed to prevent over-expression of their effects. Critically, they interact as reinforcing agents or as antagonists in a feedback mechanism.

The notion of using exogenous chemicals to control plant growth, and hence productivity, is not new. Many of the first commercial herbicides (auxin herbicides) were developed as a direct consequence of plant growth regulator research (see Chapter 1, Section 3.3).

Plant growth regulators (PGRs) are agents that interfere with the normal processes governing plant and crop development, thereby modifying yield quantity and/or quality. PGRs cannot raise the yield potential of a plant beyond its genetically determined threshold but act to fine tune its response to environmental change, permitting commercial yields to be maintained in otherwise sub-optimal circumstances. Thus, in apples, thinning agents are used to reduce the proportion of small and unsaleable fruit, allowing photosynthate to be utilised by the remainder and ensuring the production of high quality, commercially useful produce but without an increase in total weight of fruit. Similarly, losses in cereals due to lodging and the subsequent mechanical problems encountered in harvesting, and directly through fungal attack, can be avoided by the use of growth inhibitory agents that increase the standing power of the crop.

2 PLANT HORMONES

A comprehensive description of the natural plant hormones is beyond the scope of this chapter but several reviews are available. However, since the synthetic PGRs mimic or antagonise the action of plant hormones, a brief summary of their role is given.

2.1 Auxins

The first chemical shown to be active in a biological assay for plant growth substances was auxin A, 44 mg of which was extracted from a surprising starting material of 150 litres of human urine. Although similar compounds were characterised from this source, later attempts proved unsuccessful and the discovery appears to be another example of the role of serendipity in the advance of agrochemical research. β-indole-acetic acid or IAA (Figure 5.1) is now known as auxin and is an example of the many similar substances found since the early 1930s in a series of legendary plant physiology experiments.

Auxins are implicated in many developmental processes in plants,

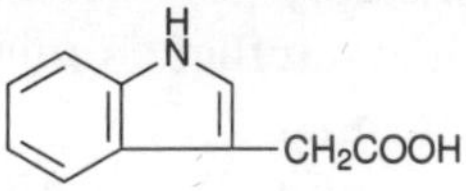

Figure 5.1 *β-indole-acetic acid*

including shoot growth, root growth, bud initiation and abscission of plant parts. They are active at very low concentration and the effects are qualitatively and quantitatively concentration dependent. Thus at up to 10 ppm of exogenously applied auxin stem growth is stimulated but beyond that concentration extension growth is retarded. In roots, much lower concentrations are required to elicit a positive growth response.

Auxins produce many changes in cells, including increased wall plasticity, enhanced respiratory rate and alterations in nucleic acid metabolism. Although the primary mode of action of auxins is not fully understood the development of the chemistry has progressed rapidly. Early discoveries identified and exploited the herbicide potential of super-optimal concentrations of auxin (Chapter 2, Herbicides) but their

PGR potential has been limited largely to root inducing-chemicals and applications to control fruit quality.

2.2 Gibberellins

Gibberellins (Figure 5.2) also stimulate extension growth but in contrast to auxins have no effect on roots. Their effect on extension growth, however, is profound, a lack of gibberellin being associated with dwarfness in plants. Varieties bred specifically for that trait are deficient in gibberellin and can be 'normalised' by gibberellin treatment. Similarly, plants that grow as prostrate rosettes can be made to shoot upwards by the application of gibberellin. However, gibberellins only affect the form of the plant, not the number of leaves or yield.

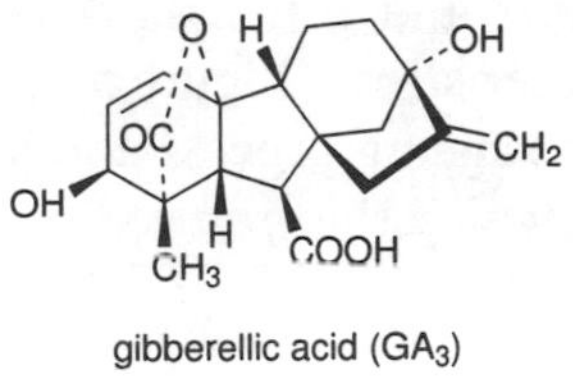

gibberellic acid (GA₃)

Figure 5.2 *Gibberellin A₃*

2.3 Cytokinins

Cytokinins were discovered in the 1950s. In combination with auxin, they control cell division, promote juvenility or slow ageing and induce the formation of lateral buds (Figure 5.3).

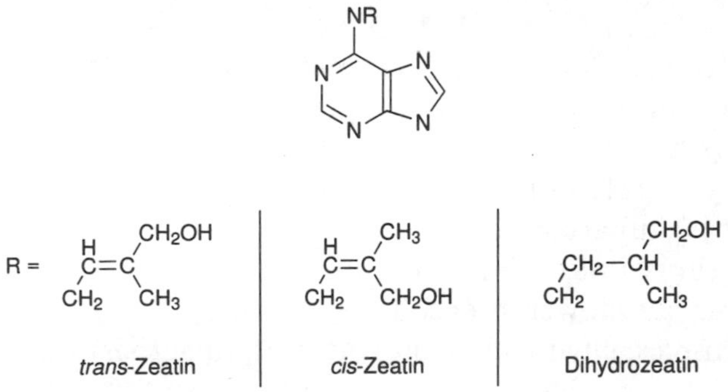

Figure 5.3 *Naturally occurring cytokinins*

2.4 Ethylene

The discovery of ethylene as a plant growth substance followed from the observation in Germany that leakage of gas used in street lighting was associated with the defoliation of trees. Subsequent research demonstrated the active component as ethylene, which was then implicated in the endogenous acceleration of ripening, the promotion of flowering and sex determination in addition to abscission of plant parts.

Ethylene action depends on binding to a receptor site that has a low affinity for other olefines. Oxygen is required for the response, which is competitively inhibited by carbon dioxide.

2.5 Abscisic Acid

Abscisic acid, as the name suggests, has been implicated in the control of abscission of leaves, flowers and fruits, as well as with the function of stomata in response to water stress (Figure 5.4). Abscission involves the synthesis of cellulase in the ageing process and it is thought that abscisic acid influences the rate at which this proceeds.

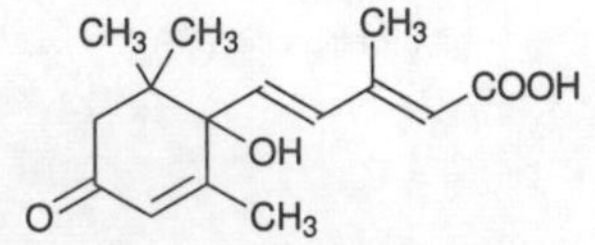

Figure 5.4 *Abscisic acid*

3 TARGETS FOR PLANT GROWTH REGULATORS

The commercial value of a crop can be measured in several ways. Man utilises all parts of plants, depending upon the species, hence value is associated with the production of:

- seeds (cereals, legumes, oil-seed rape, cotton, maize);
- roots (sugar beet, carrots, turnips);
- stems (sugar cane), tubers (potatoes);
- axillary buds (sprouts);
- petioles (rhubarb, celery);
- leaves (brassicas, onions, horticulture);
- flowers (cauliflower, broccoli, horticulture);
- fruit (apples, citrus, bananas, soft fruit, tomatoes).

The aim of PGR research is to define agents that act directly to increase the yield value of a crop. Yield is measured as quantity or quality, or both. The value of an apple crop, for example, is first dependent upon quality, in contrast to sugar beet in which yield, that is sugar content, is quantitative. However, the balance between the accumulation of assimilate as yield and its diversion to non-yield material is an equally important feature in productivity. It is economically undesirable to produce a high quality yield if, at the same time, a disproportionate amount of useless material is also produced. For example, the addition of high levels of nitrogen to a growing crop may selectively favour the production of foliage at the expense of grain production. Predictably, to be able to modify the growth and development of crops using exogenously applied chemicals is the most difficult goal in pesticide science.

The evolutionary process has selected those plant species that are able to compensate for changes in their environment. Individuals within a species may respond differently. For example, some seeds of *Avena fatua* germinate at a lower temperature than others even though they arise from the same mother plant, thus ensuring species survival under a wide range of temperature conditions. Crop plants are unnatural, being a result of man's manipulation of the available pool of characters within a particular range of species, but act in a similar fashion. They tend to be less flexible in their responses to environment but, nevertheless, exhibit some compensatory characteristics. This capacity for change is both the foundation of plant growth regulator research and the source of its limitations. Several target processes have been investigated as means to influence yield.

3.1 Control of Assimilate Production

Photosynthesis is the major means of dry matter accumulation in plants. Net photosynthesis is a measure of the gain in assimilate by the plant, after photorespiration and dark respiration losses have been accounted for, and is proportional to quantitative yield production.[1]

The control of net photosynthesis by PGRs was thought to be a potential means to enhance plant yield through the manipulation of the mechanism of assimilate gain, that is gross photosynthesis, and the mechanisms of assimilate loss, that is respiration. The inhibition of photorespiration, in particular, has been extensively researched because, under high irradiance, respiration losses from tricarboxylic acid cycle activity in C3 plants are less significant than those arising from the production and oxidation of glycolic acid in photorespiration[2] (Figure 5.5).

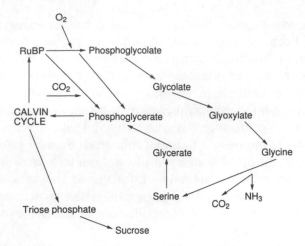

Figure 5.5 *Photorespiration*

The enzyme responsible for the fixation of CO_2, ribulose bisphosphate carboxylase/oxygenase (RUBISCO), also catalyses the oxygenation of ribulose bisphosphate to form one molecule of phosphoglycerate and one molecule of phosphoglycolate. Under normal CO_2 and O_2 concentrations six ribulose bisphosphate molecules react with oxygen for every 15 that react with carbon dioxide.

Several chemicals are known to inhibit photorespiration, the most studied being α-hydroxy-2-pyridine methane sulfonic acid (HPMS; Figure 5.6). Early work using leaf discs showed that HPMS caused the accumulation of glycolate in the light and increased carbon fixation at elevated temperatures.

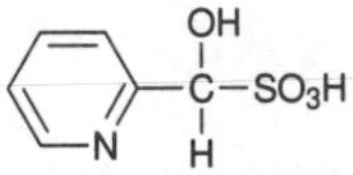

Figure 5.6 *α-Hydroxy-2-pyridine methane sulfonic acid*

This observation raised the possibility that yield could be manipulated by the use of photorespiration inhibitors. However, the results were conflicting and subsequent investigations highlighted the need to control the oxygenase activity of RUBISCO independently of the carboxylase. The high rate of turnover of this enzyme and its abundance in green tissue, together with the competitive nature of the oxygen and carbon dioxide binding severely reduces the value of photorespiration inhibition as a target for yield enhancement.

Another strategy in the manipulation of production is to consider the whole crop as an engine for assimilation. Fundamentally, yield is dependent upon the crop intercepting as much light energy as possible during the growing season. The yield of sugar beet, for example, is known to be proportional to the leaf area index of the crop (total leaf area/ground area) and the speed at which total ground cover is achieved. The role of the PGR in this circumstance is to manipulate the whole crop and not the individual plant. In soybeans with an indeterminate growth habit, self-shading of lower leaves by new, and less productive growth, can lead to yield limitation. The target for a PGR in that case would be to alter the architecture of the crop, allowing a greater interception of light by the lower leaves.

3.2 Control of Water Stress

Water stress reduces the rate of photosynthesis in plants. In soybean crops, water stress leads to a loss of root integrity. The failure of the lower leaves to intercept light in a mature crop is compounded by the water-related limitations. The result is that lower leaves abscise and, because seed pods (yield) are supplied with assimilate mainly from leaves at the same nodal position on the stem, they in turn are lost and overall yield is reduced. The control of water stress in crops is a major PGR target in yield maintenance. The alleviation of water stress in developing crops, particularly in arid areas, is also desirable.

Most terrestrial plants experience daily periods of water stress that are detrimental to growth and development. Where water demand outstrips supply the type of crops able to be cultivated is severely limited. Therefore, there is a considerable need to devise methods to reduce water loss from crops by transpiration.

The physiology of stomatal movement has been extensively researched. Early studies using phenylmercuric acetate demonstrated that the induced closure of stomata could conserve water significantly up to one month from the time of treatment. Although reductions in transpiration would probably carry a yield penalty, the long-term benefits of increased land use are desirable. However, no commercially viable anti-transpirants have been produced.

3.3 Maintenance of Crop Structure

Mature annual crops can be fragile. High yield often means that stems have to support a large weight (as seed or fruit). The larger the yield, particularly if associated with a long stem, the greater is the instability in the crop. Cereals, especially, are prone to collapse (lodging) in late season following wet and windy weather. Some leguminous (*e.g.* soybeans) and brassicas (*e.g.* oil-seed rape) plants can overgrow their neighbours in crops and encourage similar crop collapse. The subsequent inability of harvesting machinery to collect the flattened crop leads to significant yield reductions. PGRs are used to enhance stem strength and thereby prevent lodging. This is a major commercial use for PGRs, particularly the anti-gibberellins or 'retardants'.

3.4 Uniformity of Harvest

Many plants produce their yield in a stepwise fashion. This trait is most clear in crops with a continuous growth or cropping habit, for example, legumes, but it can also occur in determinate crops, for example, cereals, where tillers or similar tiers of yield-bearing side shoots are produced. Two difficulties are at once apparent. The first is practical and relates to the cost of taking harvests from a crop several times in one season. Here the ideal procedure would be to harvest all the potential yield in the one operation. The second concerns the loss of dry matter that sub-optimal yield structures represent. In cereals, the third and fourth generation tillers are less productive than the first two and do not contribute to useful yield, although in total dry weight they may comprise a significant part of the whole plant. The goal in this case is either to remove the tillers completely or to standardise them with the main tillers.

Cropping can be made uniform in some crops through the use of 'herbicidal' PGRs. The most important of these are used in sugar cane (glyphosine) to standardise ripening, measured as sugar content, and in citrus used for juicing, where ripening is made uniform through the application of PGRs that damage the fruit rind, thereby promoting the release of ethylene, a plant growth substance implicated in the onset of fruit maturity.

The time of flowering controls harvest. In some apple varieties, biennial fruiting is a feature that limits productivity. Annual cropping makes better economic sense and PGRs that retard shoot growth are used to promote yearly flowering and fruiting. In ornamentals, day-length may control flowering. This is an important consideration in the production of flowers for the winter markets, for example, poinsettias and chrysanthemums that are commonly available at Christmas in northern latitudes. Although no PGRs are available that control day-length responses, the physical manipulation of photoperiod often has undesirable side-effects such as elongation of stems. These are counter-acted by PGRs designed to retard stem growth.

The problem of tiller standardisation is essentially an exercise in the removal of the dominance that main shoots have over subsequently developing shoots. This is a characteristic that is driven by survival mechanisms; most energy is expended on the production of seed from a single shoot but others are produced as potential substitutes should the first attempt fail. Evolution has determined that such shoots are smaller sinks for assimilate than the primary shoot until the loss of that shoot occurs, when the next oldest takes over as the dominant structure. The effects of dominance can also be seen in individual cereal ears, where grain size is controlled by position.

In tobacco, side shooting is a disadvantage, acceptable yield in this crop being dependent upon the production and maintenance of large leaves. PGRs are used selectively to remove developing axillary shoots.

3.5 Harvest Quality

Harvest quality in fruit is an important factor in the marketplace. It can be measured as fruit size and/or individual fruit quality. In fruit, size is controlled largely by the presence or absence of seeds. These determine the gibberellin content of the developing fruit; the more gibberellin, the larger the fruit. Unfortunately, the general public prefers fruit that is seedless, for example, grapes, so size-quality is adjusted by the use of exogenous gibberellin treatment. Since gibberellin also controls extension growth, treated seedless grapes are unusually elongated compared with the round, seeded variety.

3.6 Other Uses

Other applications for PGRs include the enhancement of crop establishment through the promotion of roots from cuttings, the prevention of sprouting in storage potatoes and the stimulation of germination and hence of starch hydrolysis in malting barley prior to fermentation.

4 COMMERCIAL PLANT GROWTH REGULATORS

The most commercially successful PGRs are those that operate either through the inhibition of gibberellin biosynthesis, from mevalonic acid (Knee, 1982) or through the production of ethylene. These represent the two most valuable pathways for growth modification since they have key roles in extension growth, ripening, fruit set and dominance.

They act by altering the endogenous balance of gibberellins (inhibition) and ethylene (enhancement) and are often used in mixtures. Thus, ethephon, an ethylene generator, is applied in combination with mepiquat chloride, a gibberellin biosynthesis inhibitor, to enhance the overall effect.

4.1 Gibberellin Biosynthesis

Several products are available that block the production of gibberellin biosynthesis and hence control extension height. These are the plant growth retardants (Figure 5.7). They have application mainly in cereal crops in which they thicken the stem base, increase the number of vascular bundles and increase stem strength and resistance to lodging. In turn, this effect allows the grower to apply greater amounts than normal of nitrogenous fertiliser, thereby increasing yield.

In some studies, retardants have been implicated in yield enhancement mediated by the manipulation of crop canopy structure and the removal of dominance characteristics, leading to a more uniform crop and potentially higher yields. Senescence may be delayed. Although the maintenance of green tissue is a clear advantage with respect to yield production, prolonged seasons can expose the crop to damaging episodes of adverse weather and counteract the potential benefits of PGR treatment.

Some compounds, for example, ancymidol are used in the horticulture industry to induce dwarfism in pot plants.

Inhibitors of gibberellin-mediated growth and development act at several points in the biosynthesis pathway, depending upon their chemistry. The group is chemically extremely diverse (Figure 5.8) and representatives find major uses in cereals [chlormequat, ethyl 4-cyclo-propyl(hydroxy)methylene-3,5-dioxocyclohexanecarboxylate, mepiquat chloride, prohexadione], oil-seed rape (triapenthenol), ornamentals (piproctanyl, paclobutrazol, mefluidide, dikegulac, daminozide, unico-nazole), fruit (paclobutrazol, daminozide), cotton (mepiquat chloride), sugar cane (mefluidide), trees (maleic hydrazide, flurprimidol), rice (inabenfide) and turf (maleic hydrazide, flurprimidol, dikegulac). The most successful are the cationic compounds, exemplified by chlorme-quat, which inhibits the production of *ent*-kaurenoic acid from *ent*-kaurene. Perhaps to be expected is the often reported retardant action of fungicidal sterol biosynthesis inhibitors which may confer yield benefits directly and in the absence of disease by affecting the rate of leaf senescence and hence net assimilation.

No compounds increase the production of gibberellin. However, exogenously applied gibberellins are used in the malting process to enhance sugar content through the stimulation of seed germination and in the production of seedless table grapes.

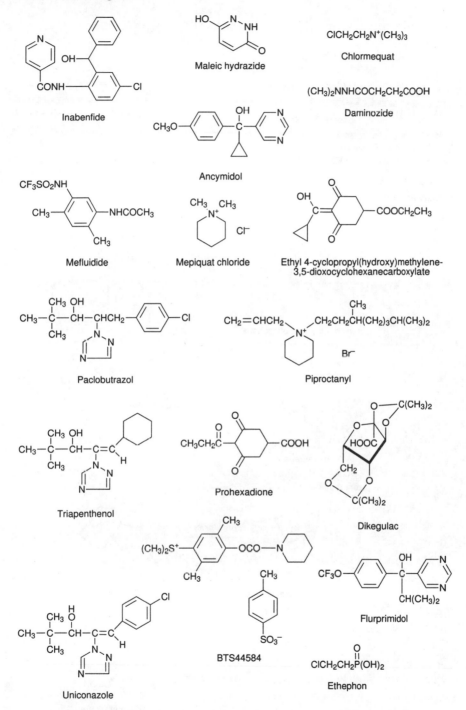

Figure 5.7 *Plant growth retardants*

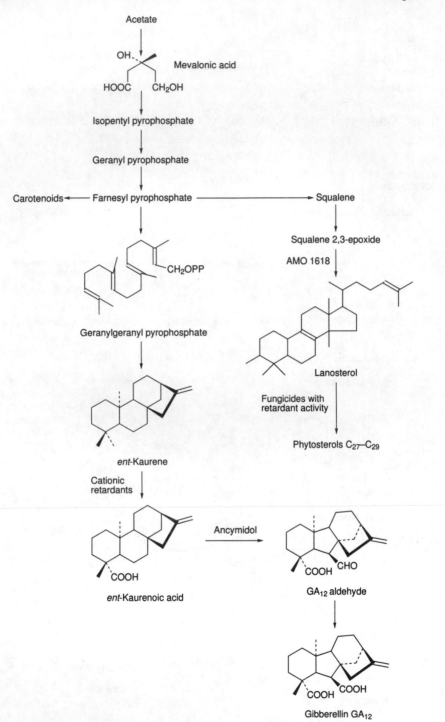

Figure 5.8 *Gibberellin biosynthesis and inhibition by retardants*

4.2 Ethylene Biosynthesis

The pathway of ethylene biosynthesis in higher plants is from L-methionine[4] (Figure 5.9). Methionine is an intermediate in other metabolic processes and the control of ethylene biosynthesis via the interference of methionine production is not realistic. The ACC synthase step from *S*-adenosyl methionine to ACC appears more susceptible to chemical modification; auxin promotes ethylene production by increasing the activity of ACC synthase. Subsequent steps from ACC are less controlled and ethylene is readily produced from the conversion of ACC in most tissues.

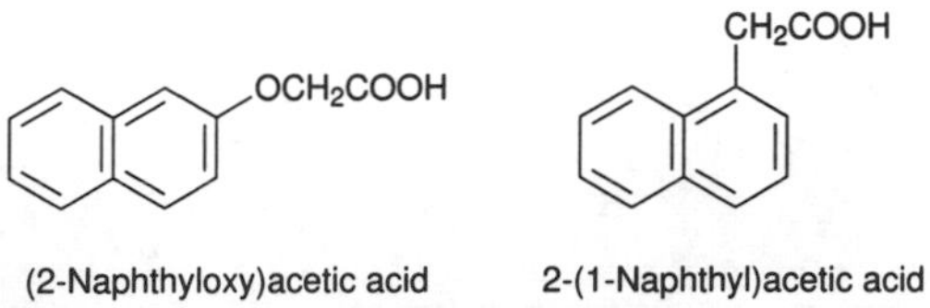
OCH2COOH
CH2COOH
(2-Naphthyloxy)acetic acid
2-(1-Naphthyl)acetic acid

Figure 5.9 *Ethylene biosynthesis*

Retardants that operate through enhanced ethylene production do not promote biosynthesis but act via chemical decomposition. Thus, at a pH of 3.5 and above, ethephon breaks down to form ethylene (Figure 5.10). Such materials are effective growth retardants and are used mainly in cereals.

ClCH2CH2—P—O⁻ + OH⁻
ClCH2CH2—P—O⁻
Cl⁻ + CH2=CH2 + HPO3 2–

Figure 5.10 *Ethylene generation from ethephon*

The stimulatory effect of auxin on ethylene biosynthesis is the basis for the use of synthetic auxins (Figure 5.11) to promote the abscission of

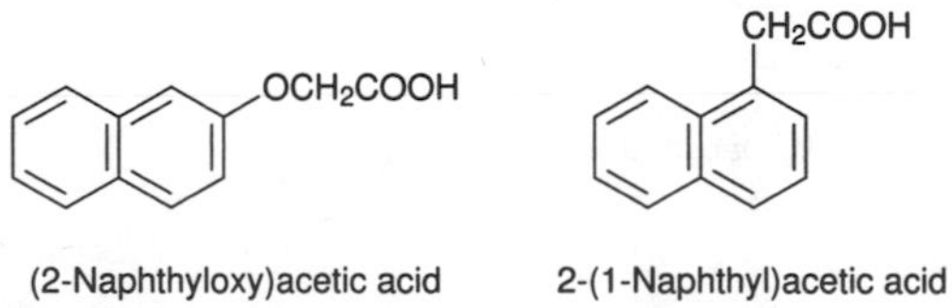
OCH2COOH
CH2COOH
(2-Naphthyloxy)acetic acid
2-(1-Naphthyl)acetic acid

Figure 5.11 *Synthetic auxins*

Sucker/side shoot control agents

$$CH_3(CH_2)_9OH$$

Decan-1-ol

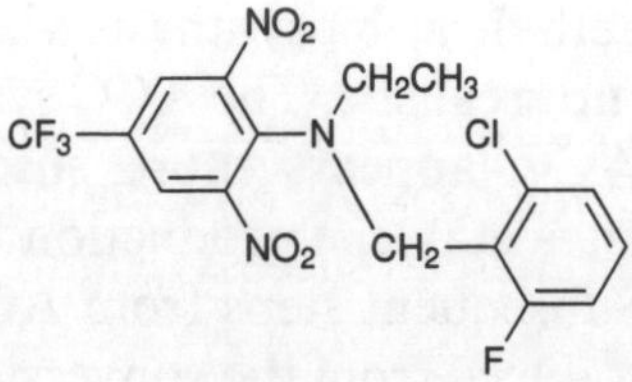

Flumetralin

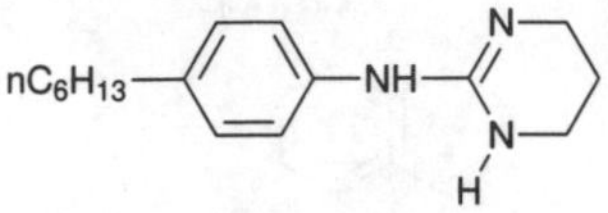

DU312098
2-(4-*n*-hexylphenylamino)-1,4,5,6-tetrahydropyrimidine

Compounds stimulating root formation

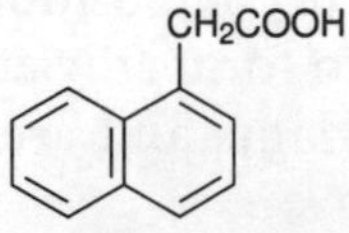

2-(1-Naphthyl)acetic acid

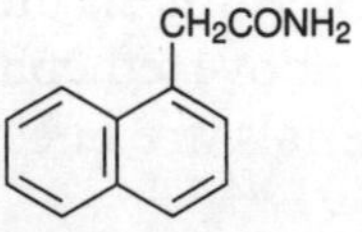

2-(1-Naphthyl)acetamide

Compounds used in fruit thinning

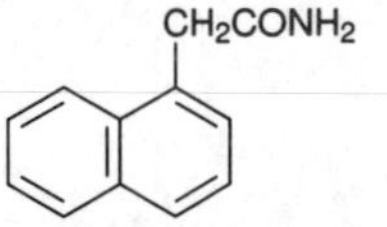

2-(1-Naphthyl)acetamide

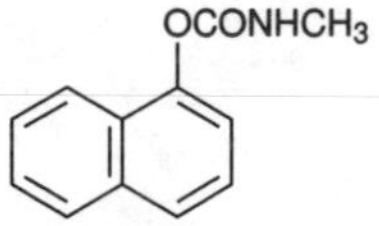

Carbaryl (see also 'Insecticides')

Compounds used in fruit setting

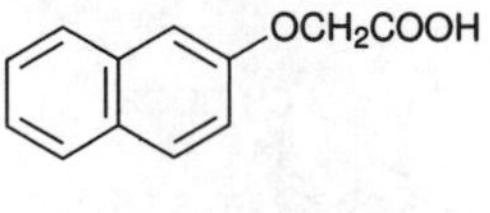

(2-Naphthyloxy)acetic acid

Figure 5.12 *Sucker/side shoot control agents*

small, low-value fruit. The remaining fruit then constitutes a greater sink for available assimilate and potentially a high-value yield.

4.3 Other Plant Growth Regulators

There is a wide variety of chemistry known to be active in the manipulation of other development processes in crops. A few examples are given but none has achieved significant commercial success (Figure 5.12).

5 QUESTIONS

5.1 Why is the plant growth regulator market the least commercially important of the four major pesticide groups?

5.2 Why does the control of water loss carry a potential yield penalty?

5.3 Can plant growth regulators increase yield?

6 REFERENCES

1. K.J. Treharne, 'Hormonal Control of Photosynthesis and Assimilate Distribution', in 'Chemical Manipulation of Crop Growth and Development' ed. J.S. Mclaren, Butterworth Scientific, 1982, pp. 55–66.
2. A.J. Keys, I.F. Bird and M.J. Cornelius, 'Possible Use of Chemicals for the Control of Photorespiration', in 'Chemical Manipulation of Crop Growth and Development' ed. J.S. Mclaren, Butterworth Scientific, 1982, pp. 39–53.
3. H.G. Hewitt, J.F. Garrod, L.G. Copping and D. Greenwood, 'The Effect of BTS44584, a Ternary Sulphonium Growth Retardant, on net Photosynthesis and Yield in Soyabeans', in 'Chemical Manipulation of Crop Growth and Development' ed. J.S. Mclaren, Butterworth Scientific, 1982, pp. 221–235.
4. F.B. Abeles, 'Ethylene in Plant Biology', Academic Press, 1973.
5. J.W. Dicks, 'Modes of Action of Growth Retardants', 'British Plant Growth Regulator Group', Monograph no. 4, 1979, pp. 1–14.

Appendix: Answers to Questions

CHAPTER 1

1.1 Reduction of loss of crop yield through direct competition, contamination by foreign bodies (insect debris or fungal mycotoxins), loss of efficiency in harvest. Their use also ensures food quality, maintains availability of low cost and wide choice of food, control of disease carrying insects, reduced energy inputs and increase biodiversity (according to Denis Avery, who argues that maximising production on the best land will release more marginal land for non-agricultural use) and guarantees we have something to eat.

1.2 The major crops for agrochemical use are cotton, rice, maize, vegetables and top fruit for insecticides; small grain cereals, rice, vines and top fruit for fungicides, and maize, soybeans, small grain cereals, rice, industrial weed control, plantations and orchards for herbicides. Other crops that may be of interest include sugar beet, oil-seed rape, potatoes and citrus dependent upon your company's presence in these crops.

1.3 The key problems for subsistence farming include pests and diseases of crops such as sorghum, chickpeas, cassava, rice, small grain cereals, maize, vegetables and sweet potatoes. Key pests include parasitic weeds such as *Striga* spp. and *Orabanche* spp. and locusts and storage diseases and insects as well as rats and mice. Snails are a problem in terms of human endoparasites as are insect disease vectors (malaria and sleeping sickness). Work would be carried out, primarily, in African and Asian countries.

1.4 Rice, $3.9 billion in 1995, cereals ($4.3 billion) represents several crops although these are primarily wheat and barley. Maize ($3.2 billion) and soybean ($3.4 billion) are also substantial markets.

1.5 Herbicides with 48% of the total agrochemical market in 1995, equivalent to $14 billion.

1.6 If compounds are to be applied as foliar sprays to control pests, diseases and weeds they will be applied to growing crops. If they are photo-unstable their persistence in the environment will be short and their biological effect short-lived. Persistent compounds will cause problems but those that are lost rapidly will not be effective. Similarly, if a compound is very volatile it will be lost to the atmosphere following application. This will reduce the effectiveness and cause environmental pollution. If it is a herbicide that is volatile, then it could cause damage to closely planted susceptible crops. (In Australia, it is not permitted for farmers to apply hormone herbicides in cotton growing areas of New South Wales in case of vapour phase movement and damage to the highly susceptible cotton crop.) If it is a fungicide or an insecticide, some volatility is often a useful property as the movement within the vapour phase within the crop canopy can lead to redistribution to new growth and better disease or insect control. Many soil applied compounds (herbicides, insecticides and fungicides) will move through the soil in the vapour phase, thereby redistributing in the crop's root zone. Such compounds are usually applied as granules or incorporated to prevent loss to the atmosphere.

See: D.T. Avery, 'Saving the Planet with Pesticides and Plastic', The Hudson Institute, Indianapolis, USA, 1995, 432 pp.

CHAPTER 2

2.1 Weeds cause significant crop loss by direct competition for moisture, nutrients and light. In addition, they often harbour pests and diseases that can invade the crop, they may be poisonous and their seeds may reduce the quality of the grain or, if sown with the harvested seed, cause weeds to spread around the farm or between farms. In many cases, the presence of weeds in a crop at harvest will reduce the efficiency of the harvester and, thereby, increase the costs to the grower.

2.2 The move into continuous cereal production and away from mixed farming with firm rotations meant that farmers could cultivate high value crops on all available land without the need for fallow. This could not have been achieved without the use of chemical herbicides. The movement of people from the land to the urban environment during the industrial revolution and thereafter reduced the available labour for hand weeding and weed control became a real problem for the farmer. The revolution was the introduction of 2,4-D and MCPA that allowed broad-leaved weeds to be controlled selectively in the previously 'dirty' cereal crops.

2.3 Any biochemical mode of action that is plant specific is a good target for a selective herbicide. Hence, the light reaction of photosynthesis, photosynthetic pigment biosynthesis, amino acid biosynthesis, plant growth regulation and lipid biosynthesis are all good targets. Poor targets include biochemical pathways that are not peculiar to plants such as uncouplers and nucleic acid biosynthesis. Other poor targets include enzymes that are present in high quantities in plants such as RUBISCO and those that are turned over rapidly.

2.4 If a plant were killed by a herbicide simply through the inhibition of photosynthesis, it should die at the same rate and with the same symptoms of a plant deprived of the photosynthetic process by other means. This could include placing the plant in total darkness. If the herbicide were only inhibiting the process it should die at the same rate as a treated or an untreated plant left in the dark. If it does not, therefore, the inhibition of the light reaction brings about plant death through a mechanism additional to or subsequent from the inhibition of the light reaction.

2.5 RUBISCO is said to be the most abundant enzyme in the world. It is certainly very abundant in green plants. To inhibit its action a very high concentration of herbicide would have to be applied to a weed and this is likely to be impractical or uneconomic.

2.6 Many herbicides are applied to plants as pro-herbicides. This means that the compound applied is not, itself, herbicidal. This is the case with esters, amines and salts of alkanoic acids and aryloxyphenoxy-propionates. Often the hydrolysis of the ester or amine to release the free acid is slow and the plant is able to modify the structure of the molecule, thereby reducing or removing its herbicidal effect. In other cases, the herbicides are rapidly hydroxylated or dealkylated, reducing or removing the phytotoxic effects. Conjugation with naturally occurring biochemicals such as glutathione removes the herbicidal effects of compounds and the rate of conjugation can vary with levels of glutathione and the presence of the enzyme glutathione *S*-transferase. In rare cases there is a difference in the herbicide's enzyme binding site and this means that the compound will not bind and, therefore, will not inhibit. With some pre-emergence herbicides the selectivity is based upon the fact that the placement of the compound in the soil profile means that it never reaches the root zone and is not taken up by the plant.

2.7 Herbicide resistance is becoming a problem in many situations. The rush of compounds that interfere with branched chain amino acid biosynthesis has meant that a very high area of land has been treated with compounds that have the same mode of action. The application of such compounds exclusively has meant that weeds that show some or complete resistance to these compounds will be selected for and resistance will increase. A more worrying occurrence is the presence of weeds that have the ability to metabolise a range of herbicides regardless of mode of action. It is important to ensure that herbicides with different modes of action be used in combination or sequentially such that site specific resistance is avoided. Metabolic resistance can be countered through the use of synergists that inhibit the detoxifying enzymes or through the use of herbicides that are less prone to metabolism in combination products.

CHAPTER 3

3.1 Insects and mites are animals and, traditionally, the most effective compounds for their control have been compounds that interfere with nerve function. Fundamentally, all animals have the same mechanisms for detecting and transmitting nerve impulses. Nerve-active compounds that kill insects will usually have some effects on mammals.

3.2 Insect growth regulation involving such processes as water balance or ecdysis (moulting) are processes that are insect specific but are also fundamental to insect survival. Another is chitin biosynthesis, as chitin is a major component of the insect cuticle.

3.3 Traditional insecticides were derived from plants. Key amongst them are nicotine and rotenone. Increasing interest is being shown in azadirachtin from the neem tree. The classical natural insecticide pyrethrin is derived from the flowers of *Pyrethrum cinerariaefolium*, from which the major insecticide group the pyrethroids were derived.

3.4 Many insects lay their eggs on the crop towards the top of the plant and the egg hatches with the neonate larva eating its way out of the shell. In many cases the larva then penetrates the crop and will not come into contact with the insecticides sprayed onto the surface of the crop. If, however, it has to walk over a treated area then it will pick up a dose of the compound and, if it is contact active, it will succumb. In addition, if a compound is stomach active the insect must feed to acquire a toxic dose and in feeding it will damage the crop. This is very important for high value fruit and vegetable crops.

3.5 If a crop is rendered insecticidal by the insertion of a gene coding for an insect-active protein then the plant will be protected from attack throughout its life cycle. It is also possible to link the toxic protein gene to a promoter that ensures that the gene and, hence, the protein is expressed in the parts of the plant that are attacked by the insect or to a promoter that is turned on when the crop is attacked by an insect. As the toxins are expressed within the crop it is unlikely that they will have any effect on the natural enemies of the insect pests. Unfortunately, the proteins will be stomach poisons and this means that they must be consumed to affect the insect and the crop will suffer some damage. In addition, the expression of an insect-active toxin for long periods throughout the life cycle of the crop might be expected to increase the possibility of the onset of resistance to the protein.

3.6 There is much talk about the avoidance of insect resistance and there are those who believe that it is inevitable and cannot be avoided. The initial problem was associated with the application of broad-spectrum persistent compounds and hindsight shows us that this is a policy that is destined to bring resistance problems. Today it is possible to monitor for the occurrence of insects and to withhold spraying until a threshhold level is reached. Early populations can be treated with 'soft' insecticides such as predators and parasites and compounds that are non-persistent and that are specific to the pest that is to be controlled. Subsequently, combinations of compounds with different modes of action applied in combination or sequentially will reduce the possibility of the onset of resistance. The use of a transgenic crop expressing two or more toxins will also help to attack the phytophagous insects and will have no effects on the natural enemies of these insect pests. It is important to ensure that the insects are not allowed to overwinter by removing crop debris and other shelter at the end of the growing season.

3.7 There is always a school of thought that the use of unnatural chemicals to protect a crop is environmentally unsound and to be avoided at all costs. This is often a cry of the ignorant who are unaware of the very strict regulatory process that is in place to ensure that all agrochemicals are without harmful effect. The usual cry is that DDT was bad. It is true that DDT would not be registered today but, at the time, it saved millions of lives. Not using chemicals to control insects would lead to a significant loss of yield and this would have complications in terms of feeding the burgeoning world population and in the cost of our food. In addition, plants respond to attack through the production of natural defence compounds that have not been evaluated through the regulatory framework and are much more toxic than conventional insecticides. Using all the weapons at our disposal, including chemical crop protection agents, we can maintain yields, reduce costs and guarantee the safety of our food.

CHAPTER 4

4.1 No. Mancozeb is a broad-spectrum fungicide used as a protectant in many economically important crops but, because of its behaviour as a general cell toxicant, its conversion into a systemic would result in an unacceptable p.iytotoxic crop response.

4.2 Non-systemics reside on the surfaces of leaves and protect the plant from fungal invasion by forming a chemical barrier against attack. Since they do not penetrate the leaf tissue, fungal structures that are within the leaf are not affected by treatment. Systemics, however, are able to destroy established internal colonies of pathogens in addition to providing protection. By definition they are distributed through-out the plant, even to areas that are not impacted by the treatment spray, although most move only within the transpiration stream. Growers are attracted to the use of systemics in preference to non-systemics because they allow effective control treatments to be made before, during or after fungal attack, they provide a uniform level of control and because they are unaffected by weather and can remain active within the growing plant for long periods, the persistence of control is high. In comparison, non-systemics have to be applied frequently and in high volume, to achieve adequate coverage of the crop. Although systemics are more expensive than non-systemics, their technological advantages ensure that they will always threaten the non-systemic market. However, multi-site of action non-systemics play a vital role in anti-resistance strategies.

4.3 Polyoxins are inhibitors of chitin biosynthesis. Chitin is a component of fungal cell walls except in Phycomycete fungi; these contain cellulose as their major cell wall constituent, hence polyoxins are inactive against *Plasmopara*, *Pythium* and *Phytophthora*.

4.4 Fungicides with specific modes of action place a strong selective pressure on sensitive fungal populations. The main reason for resistance is a change at the active site resulting from mutation and the subsequent selection of fit, insensitive individuals.

4.5 Systemic acquired resistance relies upon the stimulation of the natural host mechanisms against fungal attack. Only fungi that are sensitive to those chemical antagonists will be controlled.

4.6 The ideal fungicide could be a synthetic material or a plant activator but would have to possess several or all of the following characteristics:

- novelty, for patent protection;
- new mode of action;

- phloem mobile, to achieve good distribution;
- persistent, for long-term control, but with acceptable residual characteristics in the harvested crop;
- active across a broad spectrum for universal use;
- selective in the crop and of low toxicity to non-target organisms for safety;
- active as a protectant, curative and eradicant but with strong protectant control to allow for flexibility of treatment timing and to extend the period of control;
- compatible with other agro-products;
- possess a low resistance risk;
- commercially competitive.

CHAPTER 5

5.1 Plant growth regulators, by definition, seek to enter and manipulate the responses of plants and crops to their environment. They interact with endogenous mechanisms of control and can themselves be regarded as a change to the normal environment of the plant. Compensatory processes to counteract change are operative in all plants, including crops. The changes in growth pattern that PGRs induce last only as long as the PGR remains active within the plant. In some cases, as in plant retardation or fruit thinning, subsequent plant development may not remove the regulator's effect and a useful outcome will be obtained. In other instances, the compensatory mechanisms in plants and crops may remove the effect completely. Since the outcome is dependent upon hormonal interactions within the plant and with the prevailing environment, the overall result is difficult to predict.

To be successful, a pesticide has to provide the grower with an acceptable result in at least 80% of treatments. PGRs have failed to achieve that level of consistency in most applications.

5.2 Carbon dioxide and water flux in plants is controlled by stomatal movement. In turn, this is governed by the concentration of carbon dioxide within the leaf tissue and the concentration gradient in water vapour between the internal and external environments of the leaf. Generally, stomata are open when fully turgid, that is when water supply is not limiting. Under those conditions, maximal rates of carbon dioxide exchange are possible. Under conditions of water stress, the stomata close and water loss is reduced to a minimum but there is also a decrease in photosynthetic rate. Anti-transpirants control water loss by inducing stomatal closure. Although in the short-term the action is not detrimental to carbon assimilation, long periods of stomatal closure result in a severe depletion of assimilate and yield.

5.3 Crop yield is ultimately under genetic control. However, it can be, and usually is, profoundly modified by the environment, by competition and by weeds, pests and diseases.

Although plant growth regulators can alter some of the factors that govern productivity, their effects cannot exceed the genetic potential of the plant. Thus it is correct in most cases to regard PGRs as tools that help to maintain yield nearer to that potential than would otherwise be possible. However, in other situations, such as in the production of quality fruit, PGRs can be regarded as agents that alter growth and development patterns beyond normality; large fruit can be obtained from plants that under even the best conditions possible would not be achieved. After all, plants produce fruit as vehicles for their genes and not for the benefit of man. It is more advantageous to survival to produce many relatively small fruit than just a few large ones. Our manipulation of fruit size strikes against this evolutionary pressure but, as before, can never exceed what is genetically possible.

Subject Index